A Laboratory Guide to Frog Anatomy

THE AUTHOR

Eli C. Minkoff (Ph.D., Harvard University) is assistant professor of biology at Bates College, Lewiston, Maine. Dr. Minkoff has published articles in the *American Naturalist* as well as other scholarly zoological journals and has presented papers at the meetings of many of America's leading scientific societies.

A Laboratory Guide to Frog Anatomy

Eli C. Minkoff
Assistant Professor of Biology, Bates College, Lewiston, Maine

Pergamon Press Inc.

New York · Toronto · Oxford · Sydney · Braunschweig

PERGAMON PRESS INC.
Maxwell House, Fairview Park, Elmsford, N.Y. 10523

PERGAMON OF CANADA LTD.
207 Queen's Quay West, Toronto 117, Ontario

PERGAMON PRESS LTD.
Headington Hill Hall, Oxford

PERGAMON PRESS (AUST.) PTY. LTD.
Rushcutters Bay, Sydney, N.S.W.

PERGAMON GmbH
Burgplatz 1, Braunschweig

Library of Congress Cataloging in Publication Data

Minkoff, Eli C.
A Laboratory guide to frog anatomy.

1. Frogs--Anatomy. I. Title.
[QL668.E2M56 1975] 597'.8 74-22206
ISBN 0-08-018315-8

Printed in the United States of America

To the memory
of my father

Contents

"THUS USE YOUR FROG:

.... Put your hook through his mouth, and out at his gills; ... and then with a fine needle and silk sew the upper part of his leg, with only one stitch, to the arming-wire of your hook; or tie the frog's leg, above the upper joint, to the armed-wire; and in so doing

USE HIM AS THOUGH YOU LOVED HIM."

IZAAC WALTON

Preface

Frogs of the genus *Rana* (including both *Rana catesbeiana* and *Rana pipiens*) are commonly used throughout the world for dissection in many high-school, college, and university courses. In introductory biology or zoology courses, the frog may be presented as a "typical vertebrate"; in more advanced courses such as those in comparative vertebrate anatomy, the frog is used to represent one of the two major groups of living Amphibia. Since this manual was designed with both needs in mind, additional instructions for the advanced dissector are included in a manner that permits the individual instructor to assign them or not as he or she sees fit, but perhaps the greatest advantage is to the inquisitive student who wishes to dissect further than the class as a whole.

The need for this manual first became apparent to me while teaching an introductory course in biology. A laboratory manual was needed for students performing their first major dissection, yet it was felt that the more inquisitive or more advanced student should be able to use his dissection manual to find structures usually not encountered in introductory laboratories. Both these goals required that the manual be well *illustrated*, and that it contain extensive *directions* rather than mere descriptions of structure. No existing manual was found which would meet all these needs, so I began writing a set of mimeographed notes for each laboratory session, and it was from these that the present manual soon developed.

Illustrations have been provided throughout, especially in connection with the skeletal and muscular systems. All the illustrations are new, and most are based on my own dissections. Illustrations have always proved useful to dissectors since the days of Vesalius, but I have also noticed a disconcerting habit on the part of some students of relying too heavily on the illustrations to the neglect of text descriptions. This is unfortunate, for when a choice is to be made between two or more neighboring structures, the descriptions in the text will usually clarify the distinctions between them better than will any illustration.

For classroom purposes, doubly injected frogs are usually adequate; triply injected frogs are more expensive and add little to the student's understanding. The chapter on the circulatory system has, however, been written for use with either doubly or triply injected frogs. For the study of the skeletal system, it is also assumed that mounted frog skeletons be available to the students. Additional material is optional, but may include microscopic slides of the various organs and tissues under dissection each time.

At Bates College, we have found it convenient to devote seven laboratory sessions in our introductory course to the frog. Chapters 1 and 2 (or 1 and 4) are covered in the first session, and one session is spent on each of the remaining five chapters. The seventh week is devoted to a laboratory practical examination on the material. In large classes, this would consist of a series of dissections with one or two structures to be identified at each station. For small classes, oral practical examinations are recommended; though more time-consuming, they are usually more revealing and more instructive.

In order to permit this manual to be used according to varying teaching needs, the major subdivisions of each chapter have been clearly marked. Instructors may therefore devote two or more laboratory sessions to any one chapter, or alter at will the sequence in which the systems are dissected (studying viscera before muscles, for example, or veins before arteries). Also, each chapter contains additional directions for more advanced dissection; the laboratory instructor is at liberty to use or omit these sections as he sees fit, or to use them in certain chapters only. As a whole, this manual has been designed with the varying needs of different classroom situations in mind. I hope that each laboratory instructor will find it capable of meeting his individual needs.

E. C. M.

Introductory Remarks

(to the student)

A. GENERAL

Frogs of the genus *Rana* are distributed across all the world's continents, except where the climate is either too dry or too cold. Both the smaller species, such as *R. pipiens* (the grass frog or leopard frog), and the larger species, such as *R. catesbeiana* (the bullfrog), are commonly dissected. The position of this last species in the classification of vertebrates may be shown as follows:

Phylum Chordata (animals with a notochord)
    Subphylum Vertebrata (animals with a backbone)
        Class Amphibia (amphibians)
            Order Anura (frogs and toads)
                Family Ranidae
                    Genus *Rana*
                        Species *Rana catesbeiana*

The frog is neither too primitive nor too advanced a vertebrate to be used to introduce students to their first major dissection. Its circulatory and urogenital systems, for example, are more primitive in their arrangement, and therefore easier to understand, than are the same systems in reptiles, birds, or mammals. Yet, the frog possesses lungs (and a pulmonary circulation) which are more likely to be familiar to the beginning student than are the gill systems of fishes. As a land vertebrate, the frog possesses many muscles which can be used to illustrate the mechanism of limb movement typical of higher vertebrates. The brain of the frog is sufficiently primitive that its component parts can be seen and understood readily.

The frog is also a suitable animal for use in more advanced courses in comparative anatomy or herpetology. Since this manual was designed with both needs in mind, many chapters contain additional instructions for the advanced dissector. Individual instructors may require or not require these sections as they see fit, but perhaps the greatest advantage is to the inquisitive student who wishes to dissect further than the class as a whole.

You will note throughout this manual that new anatomical terms are printed in **boldface** when they first appear. This permits the finding of structures in class, and it is also helpful in reviewing for laboratory practical examinations. One customary form of examination is for the student to be confronted with a series of dissected frogs, each with one or two pins inserted in or pointing to different structures; the student then must supply the name of the structure indicated by the pin. The boldface words in this manual provide a list, convenient for student and teacher alike, of structures likely to appear on such examinations.

B. DIRECTIONAL TERMS

Before you proceed any further, make sure that you understand the meanings of the following directional terms:

Cranial, Caudal:

Cranial is toward the head end of the body; caudal is toward the tail end of the body.

Anterior, Posterior:

Anterior is that direction in which locomotion usually takes place; posterior is the opposite direction. In most animals (except man), the cranial end is anterior and the caudal end is posterior; thus many zoologists will consider "anterior" a synonym of "cranial" and "posterior" a synonym of "caudal." (In human anatomy, however, "anterior" means ventral, "posterior" means dorsal, and cranial and caudal are usually called "superior" and "inferior," respectively.)

Dorsal, Ventral:

The dorsal side of the body is the back; the ventral side is the belly or underside ("front"). In all vertebrates, the spinal column (or backbone) runs along the dorsal side of the body. Four-footed animals usually stand with their ventral side closest to the ground.

Medial, Lateral:

Medial is toward the median sagittal plane of the body. (The median sagittal plane is a plane which divides the body into equal right and left halves.) Lateral is further away from this plane.

Proximal, Distal:

Proximal is toward the base or attached end of a protruding structure such as a limb. Distal is further away from the attached end.

Superficial, Deep:

Superficial is toward the surface, i.e. the nearer surface. Deep is further away from the surface. Superficial structures are overlying, deep ones are underlying.

Longitudinal, Transverse, Oblique:

Longitudinal means parallel to the major (longest) axis of the body or of any elongated structure. Transverse means at right angles to such a major axis, or parallel to a minor axis. Oblique means in a direction neither parallel nor at right angles to such a major axis.

Intermediate directions:

Intermediate directions are designated by compound terms. Thus, posterodorsal is a direction intermediate between posterior and dorsal, and if *A* is posterodorsal to *B*, it is both posterior and dorsal to *B*. Similarly formed compounds include anterodorsal, posteroventral, craniolateral, posteromedial, etc.

CHAPTER 1

Major External Features

In its general body shape, and in many of its internal features, the frog exhibits modifications associated with its peculiar method of locomotion. The **hind limbs** are enlarged and strengthened, while the **forelimbs** are small. The rest of the body forms a greatly shortened, compact, central mass. The **neck** is reduced, and the freedom of movement of the head on the body is therefore greatly restricted. The shortening of the vertebral column, including the fusion of many of its elements, also restricts greatly the freedom of movement of the central body mass.

The **head** is flattened dorsoventrally, and its general outline is rounded so as to offer less resistance to movement when the frog is in the water.

Examine the **eye**. It contains upper and lower eyelids, and, in addition, a **nictitating membrane**. The latter is a white fold of skin (transparent in life) which covers the eye itself beneath the eyelids. Pull the lower eyelid forward, and with a pair of forceps lift the nictitating membrane across the front of the eye. In life, this membrane protects the frog's eye while he is underwater, but it still permits him to see.

Locate the **tympanum**, or eardrum, just posteroventral to the eye, and slightly larger. Anterior to the eyes, and slightly more medial, locate the **external nares**, which are the openings of the nostrils to the outside.

The hind limbs are large and muscular, having been adapted for jumping on land as well as for swimming (using what swimmers call the "frog kick"). Spread apart the adjacent toes of the hind feet, and notice the extent of the **webbing** between them. The webbing aids in swimming, offering more resistance to the water during the propulsive stroke.

Compare the **manus**, or hand, with the **pes**, or hind foot. How many digits has each? The **thumb** (**pollex**), or first digit of the manus, is greatly reduced; in the males, it usually functions as a **nuptial pad** to aid in grasping the female during mating.

Between the hind limbs, and dorsal in position, find the **anus**, which represents the opening to the outside of the **cloaca**, or common terminal portion of the digestive, excretory, and reproductive systems.

Your frog probably has a large gash in one of its thighs. This cut was made by the supply house in order to expose the sciatic vein. Blue latex was then injected into the major posterior veins through the sciatic vein. On the ventral side of the animal, just posterior to the forelimbs, a similar, transverse incision has probably been made in order to expose the heart. The arteries have been injected with a red latex through the ventricle, and those veins not reached in the earlier injection are also injected at this time. These latex injections are helpful because they make the arteries and veins easier to distinguish from each other and from other structures.

The **skin** of the frog is an organ whose importance should not be overlooked. In life, it must always be kept moist, for it lacks other adaptations (such as the horny epidermal scales of reptiles) for retarding the loss of moisture. It is richly supplied with blood vessels, and is therefore an important organ of respiration. If the appropriate demonstration material is available, examine under the microscope prepared slides of frog skin. Note the numerous, simple **mucous glands** which aid in keeping the skin moist; also the larger and less abundant **poison glands**.

CHAPTER 2

Skeletal System

Mounted frog skeletons will be provided by your instructor. Study these together with the accompanying diagrams. Beginning students should, as a bare minimum, learn the names of all bones except those of the carpus, tarsus, and skull. Ask your instructor whether he expects you to know the individual bones of the carpus, tarsus, and skull as well. Advanced students should learn the names of all bones, and of the prominent features on each bone as well.

(Advanced students should note that although the dermal and endochondral bones of the skull are listed together, they really have quite separate phylogenetic histories. The **endochondral** skull bones, which are preformed in cartilage, are only six in number. Of these, the sphenethmoid, prootic, and exoccipital bones are part of the axial skeleton, to which the vertebrae, ribs, and sternum also belong. The articular, quadrate and mentomeckelian are endochondral bones belonging to the visceral, or gill-arch, skeleton. All the remaining skull bones are dermal. The **dermal** bones, which are not preformed in cartilage, are the last remnants of a once more extensive dermal armor such as was present in the placoderms.)

If the appropriate material is available, examine slides of the two major types of bone tissue, **compact bone** and **spongy** or **cancellous bone**. Note that compact bone contains many concentric, cylindrical lamellae arranged around a central blood vessel; each set of concentric lamellae constitutes one **Haversian system**. Spongy bone is of similar construction, but its numerous spaces are filled in life with **marrow**, a type of blood-forming tissue.

A. SKULL

The skull of the frog is flattened dorsoventrally. It is also highly specialized and degenerate: many bones normally present in other amphibians have been lost, and others greatly modified in shape. Also, a great amount of cartilage is still present even in adult frogs; most of the bones which would form within this cartilage are among those which frogs have lost.

Identify the following bones:

1. **Premaxilla**, a paired bone, forming the anteriormost portion of the upper jaw. The premaxilla has a series of small, fine teeth, similar to those on the adjoining maxilla. The premaxilla sends out a dorsal process which forms the medial side of the external nares. Take this opportunity to notice the structure of the **nasal passage**, including the **olfactory** (or nasal) **sac**, which projects posteromedially from about the middle of the nasal passage into a blind end beneath the sphenethmoid bone.
2. **Maxilla**, a paired bone, forming the major portion of the upper jaw. The maxilla has a series of small, fine **maxillary teeth**, which you can feel by inserting your finger inside your frog's mouth and withdrawing it while pressing against the roof of the mouth.
3. **Nasal**, a paired bone, which forms the anterior margin of the very large **orbit**, or eye socket, and the posterior margin of the nasal passage (or choana). The **nasal** (or **olfactory**) **capsule** is a blind pocket extending medially or posteromedially from the nasal passage; it is roofed over by part of the nasal bone.
4. **Vomer**, a paired bone, forming the floor of the nasal capsule and part of the nasal passage. Each vomer bears a very short series of small **vomerine teeth**, which are also visible on the roof of your frog's mouth (Chapter 4).
5. **Palatine**, a paired bone, best seen from the ventral side, or, if your prepared skeleton does not permit this, by looking at the anterior margin of the orbit from behind. The palatine extends laterally from the vicinity of the vomerine teeth to a contact with the maxilla. Together with the nasal, it forms the anterior margin of the orbit.
6. **Pterygoid**, a paired bone, forms the lateral margin of the orbit. At its anterior end, it is in contact with the maxilla, nasal, and palatine bones. At its posterior end, it expands into a triangle which partially encircles the orbit from behind.
7. **Quadratojugal**, a paired bone, runs just below and parallel to the pterygoid bone. It extends posteriorly beyond the pterygoid bone to the vicinity of the jaw suspension.
8. **Quadrate**, a paired cartilage, is usually not ossified. It forms part of the jaw suspension, together with the articular bone of the lower jaw.
9. **Squamosal**, a paired, T-shaped bone. The long stem of the T extends posterolaterally to act as a support for the jaw joint. The crossbar of the T extends posteriorly above the tympanic ring; anteriorly, it extends toward, but does not quite reach, the anterior end of the pterygoid bone.

10. **Prootic** (pronounced "pro-O-tic"), a paired bone, extends from the dorsal crossbar of the squamosal medially, where it makes contact with the exoccipital and the posterior end of the frontoparietal.
11. **Exoccipital**, a paired bone, forms the posterior end of the skull. Each exoccipital bears an **occipital condyle**, which articulates directly with the spinal column. Between the exoccipitals lies the **foramen magnum**, a hole in the back of the skull through which the brain continues into the spinal column.
12. **Frontoparietal**, a large, paired bone. The two frontoparietals lie close together to form the dorsal roof of the skull between the two orbits.
13. **Parasphenoid**, a large, T-shaped, unpaired bone, visible only on the ventral side of the skull. The crossbar of the T extends across the posterior portion of the skull, where it makes contact at either end with the posteromedial tip of the pterygoid bone. The long stem of the T extends along the ventral midline of the skull between the two orbits.
14. **Sphenethmoid**, an unpaired bone. The sphenethmoid is best seen in lateral view, between the anterior ends of the frontoparietal and parasphenoid bones. The sphenethmoid represents the ossified portion of the ethmoid region, which is otherwise cartilaginous.
15. **Mentomeckelian**, a paired bone, which makes up a small portion of the lower jaw at its anteromedial end, opposite the premaxilla. This bone and the next three together constitute the lower jaw, or **mandible**.
16. **Dentary**, a paired bone, which constitutes the major portion of the lower jaw. This bone receives its name from the fact that it is the only bone in the lower jaw which ever carries teeth, but in the frog it is toothless.
17. **Angulosplenial**, a paired bone, which runs along the dorsomedial margin of the dentary bone. Near its posterior end, this bone is expanded into a medially projecting **coronoid process**, which serves for muscle attachment.
18. **Articular**, a paired bone, which may or may not be ossified. The articular is confined to the posterior end of the lower jaw, where it articulates with the quadrate to form the jaw suspension.

B. VISCERAL SKELETON

The visceral skeleton consists of those bones and cartilages which once were associated with the gill arches. To these are usually added a number of subsidiary elements associated with the pharynx and its derivatives.

The mentomeckelian, articular, and quadrate elements, already described above as part of the skull, are really part of the visceral skeleton. In addition to these, the visceral skeleton includes the following:

1. The **tympanic cartilage**, a circular cartilage which supports the tympanic membrane, or eardrum. To locate the tympanic cartilage in your own frog, use the tip of your scalpel to cut carefully around the edge of the tympanic membrane and remove it. This will also expose the columella or stapes, lying on the inside of the middle ear cavity.
2. The **columella**, or **stapes**, a minute strut lying within the middle ear cavity. It reaches from the tympanic membrane medially as far as the inner ear region of the skull; in prepared skeletons, its lateral end can usually be seen within the ring of the tympanic cartilage. The columella is the only auditory ossicle present in amphibians.
3. The **hyoid cartilage**, which is usually absent in prepared skeletons. It lies in the throat region, and consists of a flattened **body** (or **corpus**), a pair of long, curved **anterior horns** (**anterior cornua**), and a pair of shorter **posterior horns** (**thyroid horns, posterior** or **thyroid cornua**). The anterior horns support the hyoid apparatus by articulating with the prootic bones of the skull; the posterior horns articulate with and support the cartilages of the larynx.
4. The **laryngeal cartilages**, which together form the framework of the voice box or **larynx** (see Chapter 4) and also support the vocal cords. There are two pairs of laryngeal cartilages: an anterior pair of **arytenoid cartilages**, and a more posterior pair of **cricoid cartilages**. Each cricoid cartilage possesses a long **pulmonary process**, extending toward the lung.
5. The **tracheal cartilages** surround the air passages and keep them open, preventing their collapse.

C. POSTCRANIAL AXIAL SKELETON

The axial skeleton is the skeleton of the body axis, to which the other components of the skeleton are attached. The axial skeleton of the frog consists of the vertebral column and sternum, all of whose elements are unpaired, and also one unpaired and two paired bones in the skull.

1. The **vertebral column** of the frog is very specialized, due to the shortening of the trunk of the body. There are only nine individual vertebrae present, plus the elongated **urostyle**, which continues the vertebral column further posteriorly. The first vertebra, or **atlas**, is specialized for articulation with the skull by means of two large **facets** or depressions, which receive the occipital condyles of the skull. The atlas is also unusual in lacking the **transverse processes** which are present on all other vertebrae. These processes, which actually

represent fused **ribs**, are strongest on the third and fourth vertebrae, where they support the attachments of large and powerful muscles; on the more posterior vertebrae, the transverse processes support weaker muscles and are therefore more slender. The ninth or **sacral** vertebra has specialized transverse processes for articulation with the ilia of the pelvic girdle.

Examine one of the more typical vertebrae, and observe that it consists of a ventral, solid portion, the **centrum**, and a larger, dorsal portion, the **neural arch**, which encloses the spinal cord (represented by a stiff metal wire or similar support in most prepared skeletons). The centra of the first seven vertebrae are **procoelous**, which means that they are concave only in front. The eighth vertebra is **amphicoelous**, with concave facets both in front and behind. The ninth or sacral vertebra is **acoelous**, with a convex facet anteriorly and a pair of small peglike processes posteriorly, for articulation with the urostyle. The neural arch of each typical vertebra possesses, in addition to the transverse spines, a dorsally projecting **neural spine** (absent on the atlas) and two pairs of articulating processes or **zygapophyses**, by which it articulates with the adjacent vertebrae. The **anterior zygapophyses** face dorsally and slightly medially; the **posterior zygapophyses** face ventrally and slightly laterally.

The vertebral column is continued beyond the sacral region by the **urostyle**, which represents several fused **caudal vertebrae**. The urostyle has a conspicuous dorsal keel, and also a hollow canal, best seen from the anterior end, into which the spinal cord continues as the **filum terminale** (see Chapter 7).

2. The **sternum**, or breastplate, is composed of four bones and cartilages, closely associated with the clavicle and coracoid of the shoulder girdle. Projecting anteriorly from the clavicles is an unpaired bone, the **omosternum**; beyond this anteriorly, the sternum expands into a simple, flat cartilage known as the **episternum**. Projecting posteriorly from the coracoid bones is an unpaired bone, the **mesosternum**; beyond this posteriorly, the sternum ends in a large, flat, bifurcated cartilage known as the **xiphisternum**. The fore-to-aft sequence is therefore: episternum (cartilage), omosternum (bone), mesosternum (bone), and xiphisternum (cartilage); the sequence is interrupted between the omosternum and mesosternum by the pectoral girdle.

D. APPENDICULAR SKELETON: PECTORAL GIRDLE AND FORELIMB

The **pectoral girdle** (or **shoulder girdle**) consists of a series of bones and cartilages which encircle (or *gird*) the body and provide support for the forelimb. In the frog, the pectoral girdle is firmly attached to the sternum. The pectoral girdle consists of the following bones and cartilages, all of them paired:

1. The **suprascapula**, a broad, flattened cartilage, dorsally located.
2. The **scapula**, a bone corresponding to the human shoulder blade.
3. The **coracoid**, which is the larger and more posterior of the paired, ventral elements.
4. The **clavicle**, a ventral element smaller than and anterior to the coracoid.

(The advanced student should be aware that the clavicle, usually considered as part of the shoulder girdle, does not really belong here. Unlike the true appendicular skeleton, which develops from mesenchyme tissue in the embryonic limb bud, the clavicle is a membrane bone, similar in its development to the dermal bones of the skull, and, like them, derived phylogenetically from the dermal armor of placoderms.)

The **forelimb** of the frog includes the following bones:

5. **Humerus**, the bone of the upper arm. The proximal end of the humerus (its **head**) fits into the glenoid cavity, formed by the scapula and coracoid bones. Along the more proximal half of its shaft, the humerus develops a ventrally projecting **deltoid ridge** for the attachment of muscles. At its distal end, the humerus possesses a rounded surface (for articulation with the radioulna), flanked on either side by small projections (known as **epicondyles**) for the attachment of muscles.
6. **Radioulna**, which represents the fusion of two bones, the **radius** and **ulna** of other land vertebrates. The proximal end of the radioulna is extended into a process, known as the **olecranon**, which reaches around the distal end of the humerus and serves for the attachment of muscles.
7. The **carpals**, or bones of the wrist. There are six carpals in the frog: a proximal row, consisting of a **radiale** (beneath the radius, on the thumb side), **intermedium**, and **ulnare** (beneath the ulna, on the outer side); and a distal row. The first and second **distal carpals** (beginning on the thumb side) are distinct; the third, fourth, and fifth distal carpals have become fused into a single bone.
8. The **metacarpals**, or bones of the hand proper. A vestigial bone, extending beyond the first distal carpal, represents the first metacarpal, corresponding to the thumb, which is no longer present. The remaining metacarpals are numbered II through V.
9. The **phalanges**, or bones of the fingers. The second and third digits possess two phalanges each; the last two digits possess three phalanges each. The phalangeal formula of the frog's hand is therefore 0.2.2.3.3, with the initial zero representing the absent thumb.

E. *APPENDICULAR SKELETON: PELVIC GIRDLE AND HIND LIMB*

The pelvic girdle of the frog consists of the following three paired bones and cartilages:

1. The **ilium**, an anterodorsal bone, is the largest part of the pelvic girdle. Each ilium possesses an extremely long, anteriorly projecting process, which parallels the urostyle and reaches as far as the transverse process of the ninth or sacral vertebra, with which it articulates.
2. **Ischium**, the bony posterior element.
3. **Pubis**, the anteroventral element, usually still cartilaginous in the frog.

The hind limb of the frog, greatly modified for jumping, consists of the following bones:

4. **Femur**, or thighbone. The proximal end, or **head**, of the femur fits into a socket or depression in the pelvic girdle. This socket is called the **acetabulum**, and is formed by all three pelvic bones together.
5. **Tibiofibula**, corresponding to the separate **tibia** and **fibula** of most land vertebrates.
6. The **tarsals**, or anklebones, corresponding to the carpals of the wrist. The two proximal tarsals are greatly elongated thus giving the hind limb an extra segment and an extra lever arm; these two are the **tibiale** or **astragalus** (usually the more slender of the two, opposite the tibia) and the **fibulare** or **calcaneum** (the stouter of the two, opposite the fibula). There are also four additional tarsals, which may represent the **centrale**, **distal tarsal 1**, **distal tarsal 2**, and **distal tarsals 3+4**, but these homologies are uncertain.
7. The **metatarsals**. In addition to the normal five metatarsals (I–V), there is also present a small, vestigial **prehallux** or **calcar** bone. This additional bone may represent an extra digit that was once present, or it may represent a sesamoid bone, formed within the connective tissue at a point of stress.
8. The **phalanges**. The phalangeal formula of the frog's foot is 2.2.3.4.3; the prehallux is not counted.

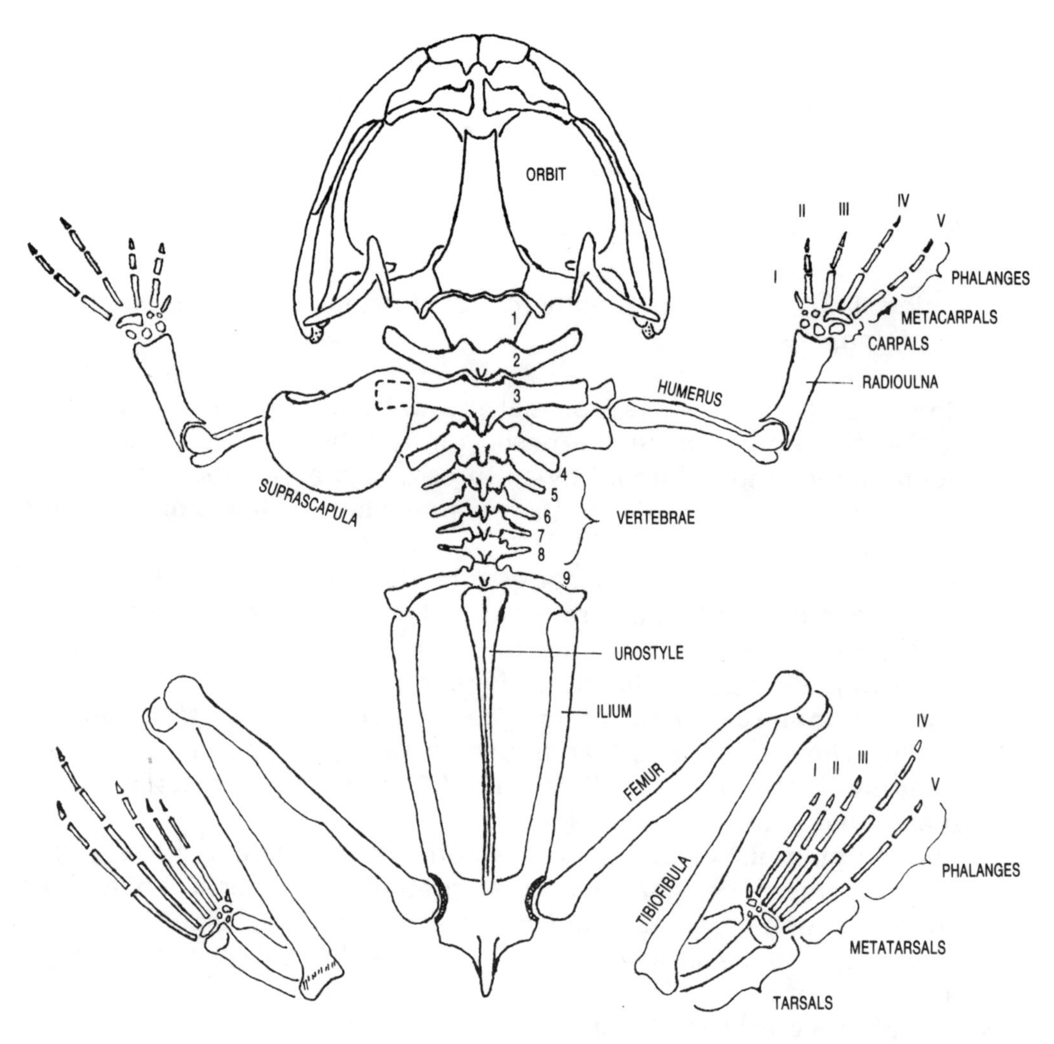

Fig. 1 The entire skeleton in dorsal view, with the right scapula (and suprascapula) removed.

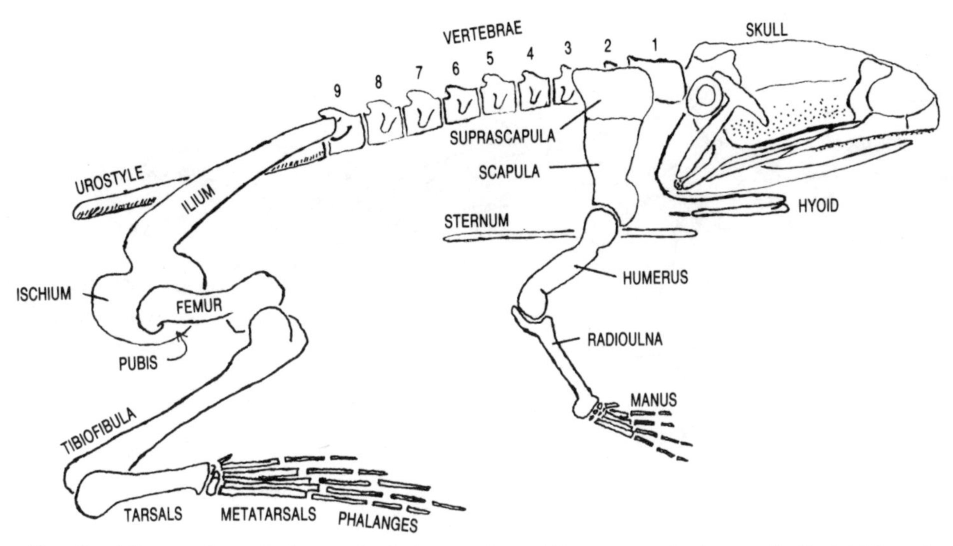

Fig. 2 The entire skeleton in lateral view. The urostyle is partially hidden by the ilium.

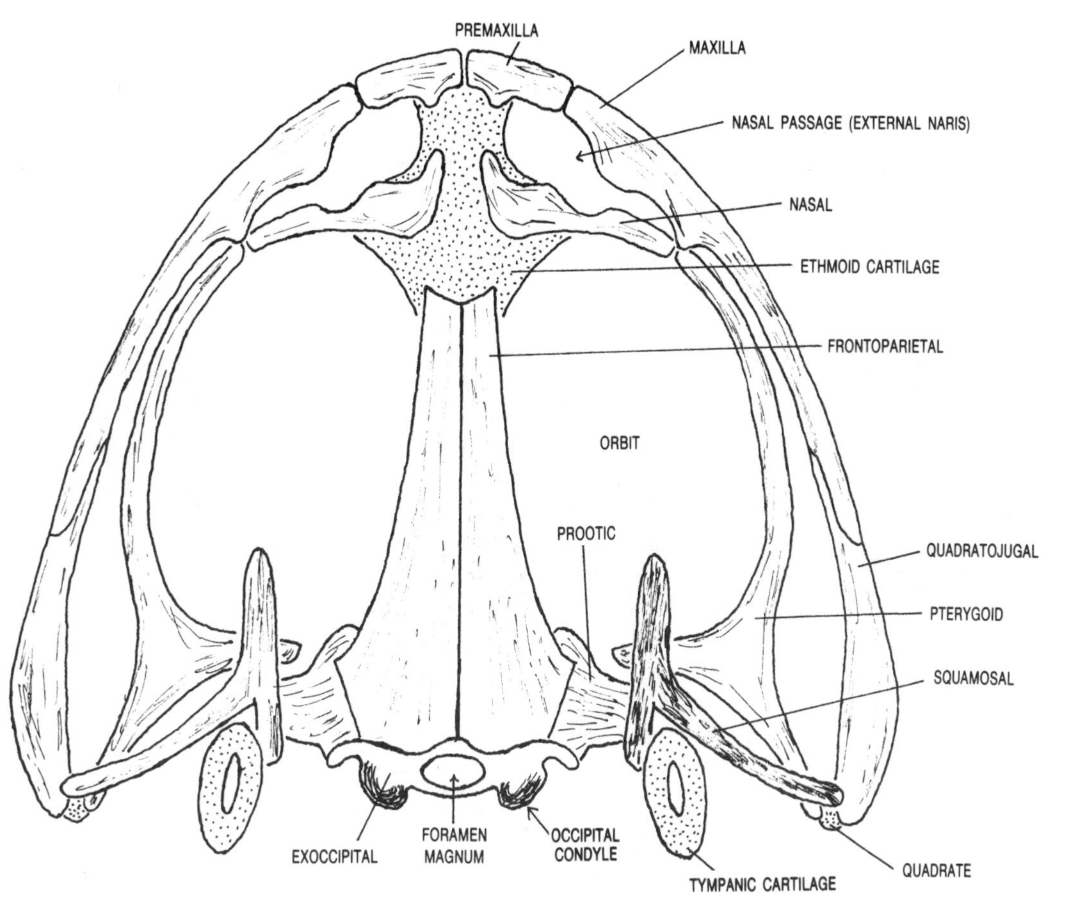

Fig. 3 Skull, dorsal view. Stippled areas represent cartilage.

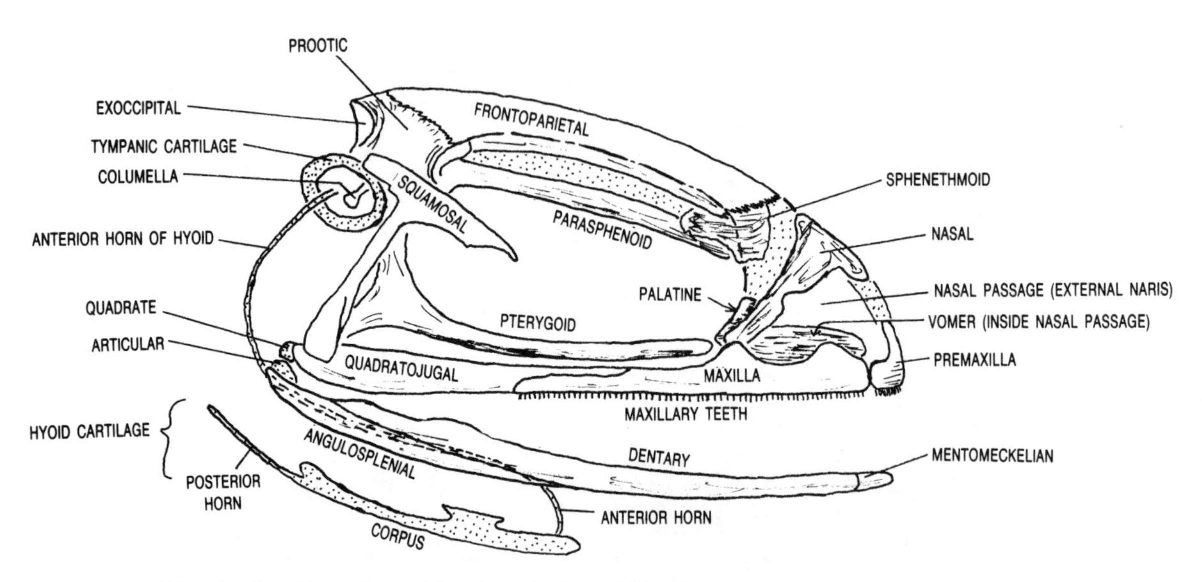

Fig. 4 Skull and hyoid, lateral view. Stippled areas represent cartilage.

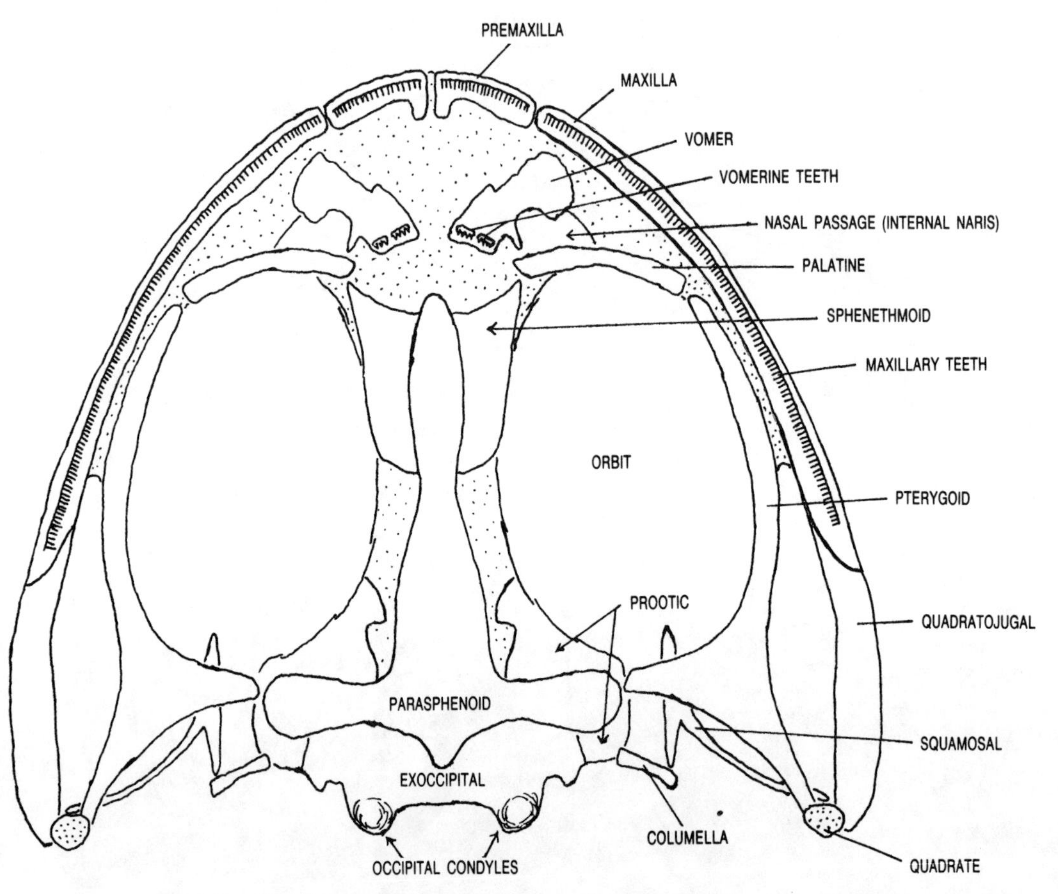

Fig. 5 Skull, ventral view. Stippled areas represent cartilage.

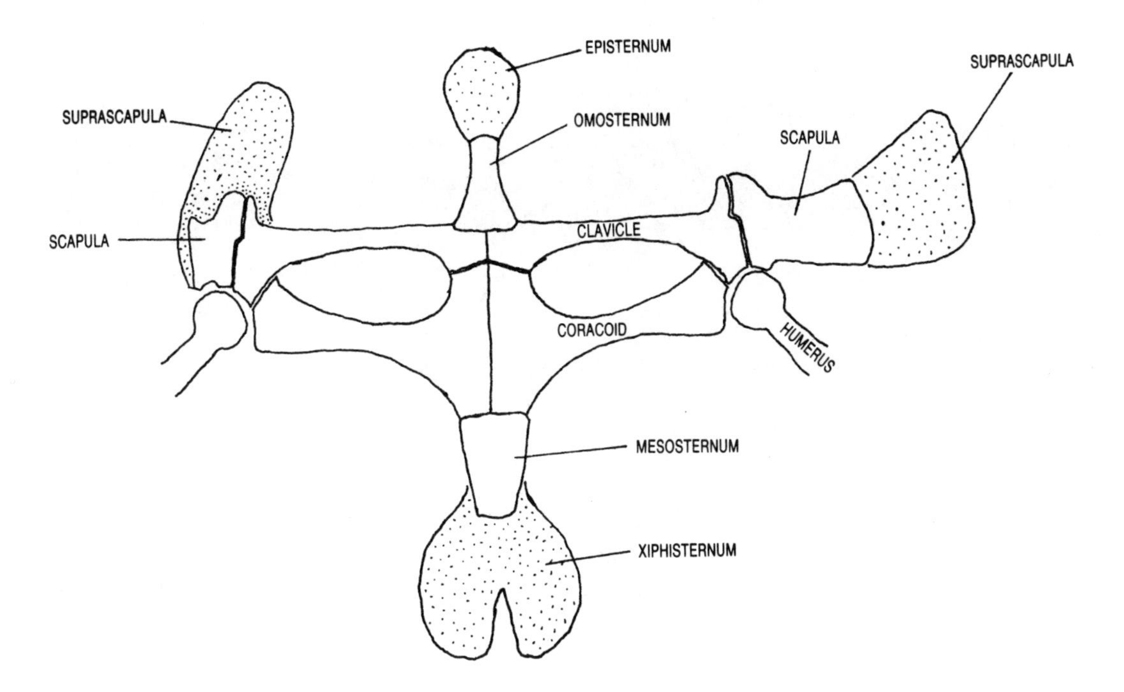

Fig. 6 Shoulder girdle and sternum, ventral view. Right side (to the reader's left) in natural position; left side with scapula and suprascapula folded ventrally into the same plane as the other elements. Stippled areas represent cartilage.

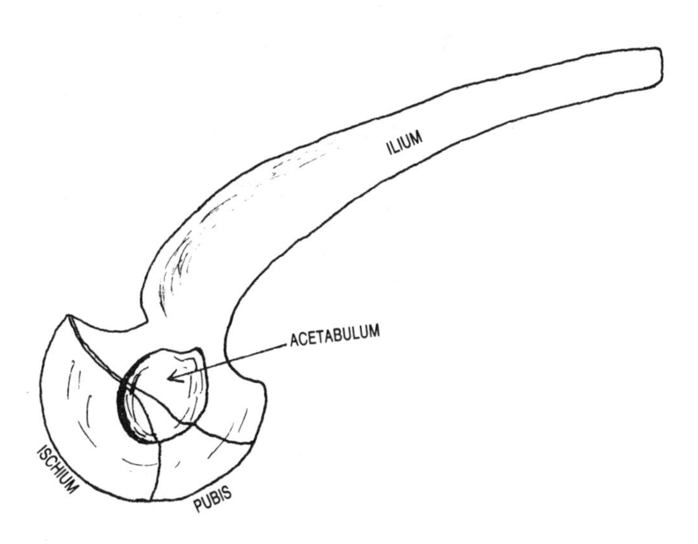

Fig. 7 Pelvic girdle, lateral view.

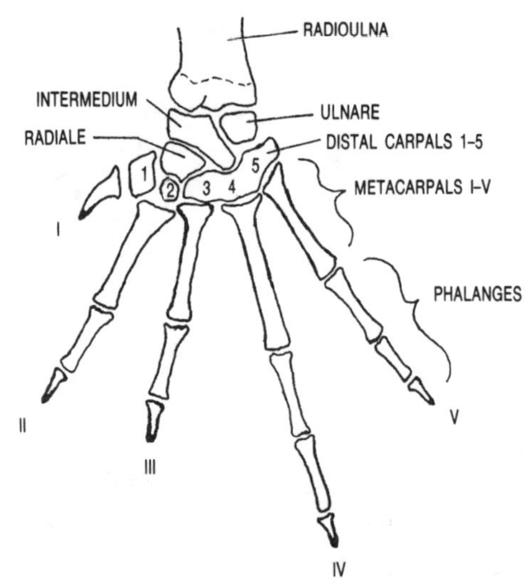

Fig. 8 Carpus and manus, palmar view.

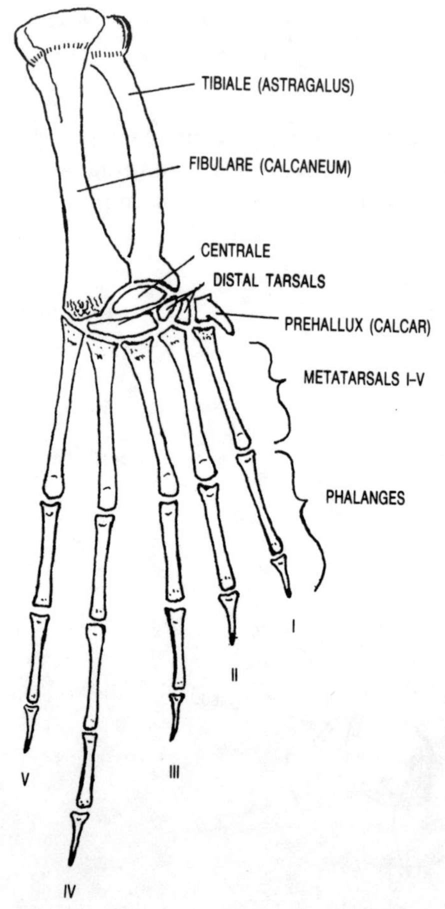

Fig. 9 Tarsus and pes, dorsal view.

CHAPTER 3

Muscular System

Please *read carefully* the following "introduction" (paragraphs A–C) *before* coming to class, and at any rate *before* you begin dissecting.

INTRODUCTION

A. MUSCLES AND THEIR STRUCTURE

Muscles are organs whose basic task is the conversion of chemical energy into mechanical energy. This energy conversion is brought about by the action of **muscle tissue**, which is a prime constituent of every muscle. Note that the term "muscle" is often used ambiguously, to refer either to an organ or a tissue. Please be careful in your own use to distinguish *muscles*, which are organs, from *muscle tissue*. Muscles are composed largely, but never entirely, of muscle tissue; connective tissue (of the "regular," dense variety) is also always present. Furthermore, muscle tissue often occurs in organs that are not muscles, such as the heart, stomach, uterus, or intestine.

The fleshy part of a muscle is called its **belly**. The belly is made up largely of parallel fibers of muscle tissue, with a sheath of dense, regular connective tissue surrounding each bundle. Sometimes the muscle tissue does not extend all the way to the attached end of a muscle, and a **tendon** is formed, consisting of the regular, dense connective tissue only. Tendons are typically long and string-shaped, but occasionally they are broad, flat, and sheetlike; such a sheetlike tendon is known as an **aponeurosis**.

The attached ends of muscles are important to the student in the recognition and identification of muscles during dissection. The fixed or proximal end of a muscle, or the end closer to the center of the body, is called its **origin**. (A muscle may have several origins; if they are distinctly separated, the muscle is described as having several **heads**.) The more remote, more distal, or more movable end of a muscle is known as its **insertion**, although the distinction between origin and insertion is occasionally an arbitrary one. If one end of a muscle is tendinous or aponeurotic, it is usually the insertion.

B. TERMS OF MOVEMENT

Before you proceed any further, make sure that you understand the following terms of movement, used in connection with muscle descriptions:

Extension, Flexion:

These are the most important and most general terms. Extension is the opening of a joint to a wider or more obtuse angle; flexion is the closing of a joint to a smaller or more acute angle. The curling of the fingers, or the bending of the elbow, are both examples of flexion; the straightening out of the fingers or the elbow are examples of extension.

Abduction, Adduction:

The bending of a structure to one side of its neutral or central position is called abduction (from Latin *ab*-, in this case meaning "away from"); the return of a structure toward its neutral or central position is called adduction (from Latin *ad*-, meaning "toward"). The spreading apart of the fingers is an example of abduction; the bringing together of the spread fingers is called adduction.

Protraction, Retraction:

A movement forward or cranially is called protraction; a movement caudally or a withdrawal is called retraction.

Types of rotational movement:

Homolateral rotation is the dextral (clockwise) rotation of the head or other median structure by a muscle on the right side, or sinistral (counterclockwise) rotation by a muscle on the left side. **Contralateral rotation** is the sinistral rotation of a median structure by a muscle on the right side, or its dextral rotation by a muscle on the left side.

Lateral rotation is the dextral rotation of a right, or sinistral rotation of a left paired structure. **Medial rotation** is the dextral rotation of a left, or sinistral rotation of a right paired structure. **Supination** is a special term for the lateral rotation of the forearm; **pronation** is a special term for the medial rotation of the forearm.

Elevation, Depression:

Elevation (or *levation*) is a movement dorsally or a raising. Depression is a movement ventrally or a lowering.

Sphinction, Dilation:

Sphinction is the closing or restricting of any opening, cavity, or hollow structure. Dilation is the enlarging of any opening, cavity, or hollow structure.

C. THE USE OF "LEVELS"

In the use of any dissection manual, the study of muscles is often the most arduous and frustrating of tasks for student and teacher alike. There are nearly 200 muscles in the frog (counting the bilaterally paired muscles only once), and it is usually not necessary for even the advanced students to learn each and every one of them. Yet, even the beginning student ought to appreciate how numerous the muscles of the body really are, even though he may be required to learn the names of only a few.

Most laboratory manuals simply omit half or more of the muscles, and the student then fails to appreciate that he is learning only a small fraction of their number. Other manuals list too many muscles, and the beginning student rightfully resents the fact that he has to learn all of these. Most frustratingly, the student may accidentally encounter a muscle not listed in his manual, and when he asks, "what muscle is *this*?", he often frustrates not only himself, but his instructor as well.

It seems that a laboratory manual should list all the muscles, or nearly all, but that the instructor should assign only a select number of these to the student. With most laboratory manuals, such selection is very bothersome to the instructor, and is therefore hardly ever made; the students are either overburdened with the learning of too many muscles, or else the study of muscles is omitted entirely. In order to make the task of selection easier for the instructor, I have indicated the *level of difficulty after the name of each muscle*, as follows:

Levels 1 and 2—For beginning students, in an introductory course where only one laboratory period or less is devoted to the study of muscles.

Levels 3, 4, and 5—For more advanced students, in an anatomy course in which two or three periods are devoted to the dissection of muscles.

Level 6—The individual muscles of the fingers, for very advanced (i.e. graduate) students only. These muscles will not be encountered accidentally in the search for other muscles, and are considered beyond the scope of this manual.

Instructors may now (indeed, they *should*) inform their students, for example, that they are required to learn the muscles of levels 1, 2 and 3 only. Since all the muscles (except those of the individual fingers) are listed, the student who accidentally finds a nonrequired muscle during dissection should be able to correctly identify it from the appropriate description.

LABORATORY EXERCISES

D. MUSCLE TISSUES

If adequate material is available, examine slides of the three basic types of muscle tissue. **Smooth muscle tissue** is the simplest type, and should be examined first. The cells of this tissue are usually long and spindle-shaped, with a nucleus in the center, surrounded by a large amount of cytoplasm. Smooth muscle tissue is involuntary, and it occurs in the muscular linings of digestive and certain urogenital organs (stomach, intestine, etc.; oviduct, uterus, etc.), in the linings of arteries, and in such other locations as the ciliary body of the eye, but never in muscles.

Cardiac muscle tissue, as its name implies, occurs only in the heart. Notice that the fibers *branch and anastomose* (rejoin, often in a different sequence) to form a single interconnected network. Notice also the **intercalated discs** separating the adjacent cells. Cardiac muscle tissue, like smooth muscle tissue, is involuntary, but in other respects it bears greater resemblance to voluntary muscle tissue: the nuclei are peripheral, and each fiber is strongly crossbanded.

Voluntary (also called **skeletal**) **muscle tissue** is characterized by the presence of numerous parallel bandings or **cross-striations**, a feature which it shares with cardiac muscle tissue. The nuclei are located on the periphery of each fiber, rather than in the center. Furthermore, many nuclei occur in the same fiber, with no cell boundaries separating them; the fiber is therefore called a **syncytium**. In the presence of cross-striations and peripheral nuclei, the voluntary muscle tissue resembles cardiac muscle tissue, but there are no intercalated discs, and the fibers neither branch nor anastomose. Unlike the other types of muscle tissue, voluntary muscle tissue is under the conscious, voluntary control of the central nervous system; indeed, it is the only tissue of the body whose actions can be controlled at will. Unlike the other two types of muscle tissue, which never occur in muscles, this tissue type is *confined* to muscles.

The voluntary muscles of the body contain connective tissue in addition to voluntary muscle tissue. The predominant connective tissue in muscles is known as **regular connective tissue** because of its parallel orientation of fibers, principally of the protein **collagen**. Regular connective tissue surrounds groups of parallel muscle fibers, arranging them into primary, secondary, tertiary, and higher order bundles. These layers of connective tissue are continuous with each other, and with the tendon or aponeurosis, if any exists. They are also continuous with the connective tissue of whatever the muscle may be attached to: the periosteum of a bone (never the bone tissue itself), the sclera of the eyeball, or the dermis of the skin. Examine a slide of a tendon, and, if possible, also of whole muscles cut in cross section (study under low power the arrangement of the muscle fibers into bundles), and of tendinous attachments to bones.

E. HOW TO DISSECT MUSCLES

Before you begin dissecting, make sure you have read the introductory paragraphs, A–C, above. You may at this time wish to reread paragraph B (terminology of movement), as well as the directional terms on pages xii–xiii of this manual.

Before dissecting the muscles in any region of the body, lift away the skin in that region by sliding the back end of your scalpel or forceps back and forth under the skin. Cut the skin with a scalpel or with scissors. Cut around (preferably with a scalpel), and leave the skin adhering to, the following regions: mouth, eye, tympanum, anus, hand (**manus**), and foot (**pes**).

With a fine forceps, clean away any connective tissue fascia which may surround each muscle. Pay particular attention to the direction of the muscle fibers; you may recognize many individual muscles in this manner. Separate the various muscles as best you can, using *blunt instruments only*. (Do not overlook the potentialities of the back end, i.e. the handle, of your scalpel or forceps as a blunt instrument.) Separate the muscles as far as their origins and insertions, *then* identify them with the aid of the descriptions and the illustrations. (Find the muscle *first*, *then* look for its matching description.)

When dissecting the muscles, refer continually to (1) the muscle descriptions, (2) the accompanying diagrams, and (3) the mounted frog skeleton. *Do not rely too heavily on illustrations for the identification of muscles*; positive identification of muscles can be made *only* by examining their origins and insertions, and comparing with the accompanying descriptions (or with a chart listing origins and insertions). If you must cut the overlying muscles in order to study the underlying ones, lift the muscle in the middle of its belly (*never* at its origin or insertion) and cut cleanly through with a sharp scalpel (not scissors). If you must cut several parallel muscles, cut each at a different length, so that you can reconstruct the cut ends without error. *Do not cut anything unless necessary*, and cut only in a manner that will allow you to piece things back together the way they were. *Use blunt tools wherever possible*, to avoid damaging other muscles.

F. MUSCLES OF THE SPINAL COLUMN, TRUNK, AND ABDOMINAL WALL

Like the greatly compacted trunk skeleton of the frog, the trunk muscles of the frog are greatly reduced, especially in size. With the foreshortening of the spinal column and the strengthening of the individual vertebrae, the spinal muscles in particular have become shorter, stouter, and reduced in both number and complexity. With the loss of the ribs, many muscles of the trunk region have been lost, but the muscles of the abdominal wall have remained more or less unspecialized. Certain spinal muscles have, as in other land vertebrates, become secondarily attached to the shoulder girdle; these include the serratus, rhomboideus, and levator scapulae muscles.

The **epaxial muscles** (*extensor muscles*) of the spinal column include the following (Nos. 1–5):

1. **M. coccygeosacralis** (level 3, Fig. 10). This muscle may be found dorsal to and between the longissimus dorsi and coccygeoiliacus muscles. Its origin is from the anterior half of the urostyle on its lateral aspect. From here, the fibers run anterolaterally, to insert upon the transverse process of the ninth or sacral vertebra. The muscle functions to fix the urostyle in place, and to move it dorsally or to either side.
2. **M. longissimus dorsi** (level 3, Fig. 10). This long, narrow muscle will be found dorsomedially, lying parallel to either side of the vertebral column, and immediately beneath the **dorsal fascia** which covers the entire epaxial musculature. The muscle fibers originate as far caudally as the anterior third of the urostyle, and insert as far cranially as the exoccipital bone of the skull. Along its length, the muscle is interrupted by several tendinous inscriptions, each of which is attached to the neural arch and transverse processes of an underlying vertebra. This muscle is the principal extensor muscle of the spinal column; the fibers of either side may act alone to abduct or adduct the spinal column dorsolaterally.
3. **M. ileolumbaris** (level 4, Fig. 10). This muscle is divisible into a *pars lateralis* and a *pars medialis*. The *pars medialis* originates largely from the transverse process of the ninth or sacral vertebra, continuing anteriorly where the m. coccygeosacralis leaves off. The *pars lateralis* has a tendinous origin from the lateral side of the ilium, from whence it runs anteriorly to join the *pars medialis* at the level of the seventh vertebra. From here, the unified muscle continues anteriorly, to insert on the transverse process of the fourth or fifth vertebra. This muscle is continued further anteriorly by portions of the m. longissimus dorsi, and, like the latter, it is interrupted several times by tendinous inscriptions anchored to the underlying vertebrae. The muscle acts as an extensor, abductor, and homolateral rotator of the back.
4. **Mm. intercrurales** (level 5, Fig. 22). These short muscles lie deep to the m. longissimus dorsi, where they run between the neural arches of adjacent vertebrae. The anteriormost of these muscles runs forward from the atlas to insert on the skull, just above the foramen magnum. The muscle is an extensor of the back.
5. **Mm. intertransversarii** (level 5, Figs. 10, 15, 22). These muscles run between the transverse processes of adjacent vertebrae as far anteriorly as the second, and from the second (not the first) vertebra to the exoccipital bone of the skull. There are two pairs of these muscles

between the skull and the second vertebra, and also between the second and third vertebrae; posterior to the third vertebra, there is only one pair of intertransversarii per segment. At various places, these muscles may be closely associated (or occasionally fused) with the deep fibers of the longer epaxial muscles, showing that all are closely related developmentally. The intertransversarii are primarily abductor muscles of the spinal column.

The remaining muscles of the spinal column and trunk are all **hypaxial muscles**, whose primitive function lay in the flexion of the vertebral column and the body as a whole. With the foreshortening of the vertebral column and the loss of ribs, many of these muscles have disappeared, but a few still remain, usually with altered functions. Of the hypaxial muscles still present in the frog, one (No. 6) is now associated with the pelvic girdle, and seven (Nos. 7–13) with the pectoral girdle. The remaining five hypaxial muscles (Nos. 14–18) have lost all connection with the vertebrae and ribs; four are associated with the ventral abdominal wall, while the last has become specialized as an anal sphincter.

6. **M. coccygeoiliacus** (level 3, Figs. 10, 18). This broad muscle lies anterior to the cloaca, on either side of the dorsal midline. Its origin is from the urostyle, along the greater part of its entire length. From here, the fibers run anterolaterally, to insert on the anterior two-thirds of the ilium. The more anterior fibers are covered medially by the coccygeosacralis muscle. The m. coccygeoiliacus functions as an adductor and retractor of the ilium; it also helps hold the urostyle in place.
7. **M. rhomboideus anterior** (level 4, Figs. 10, 15). A flat muscle, partially hidden by the suprascapula. Its origin is from the posterior end of the frontoparietal bone, near the posteromedial corner of the m. temporalis. The insertion is to the deep or ventral surface of the suprascapula near its medial border; the mm. rhomboideus posterior, levator scapulae superior, and serratus superior also insert in this region. The m. rhomboideus anterior should be cut, and the shoulder girdle folded posterolaterally, in order to expose these muscles. The m. rhomboideus anterior functions principally as a protractor of the shoulder girdle.
8. **M. rhomboideus posterior** (level 5, Figs. 10, 15). This broad, flat, triangular muscle originates from the neural spines of the third and fourth vertebrae, from the transverse process of the fourth vertebra, and from the tendinous septum that lies between these, in association with the m. longissimus dorsi. From here, the fibers converge anteriorly toward a short tendon, by which the muscle inserts on the deep or ventral surface of the suprascapula near its medial border, just

posterior to the insertion of the m. rhomboideus anterior. The primary function of this muscle is that of a retractor of the shoulder girdle; it also functions, together with the other muscles in this region, to attach the shoulder girdle to the axial skeleton.

9. **M. levator scapulae superior** (level 5, Figs. 10, 15). The origin of this muscle is from the prootic and exoccipital bones of the skull, just ventral and posterior to the ear region, and separated by the temporalis muscle from the origin of the m. rhomboideus anterior; a small number of fibers also originate directly from the tympanic cartilage. The muscle then runs posteriorly, to insert on the deep or ventral surface of the suprascapula near its medial border, just anterior to the insertion of the m. rhomboideus anterior. This muscle acts to protract the shoulder girdle, and to abduct the head toward the same side.
10. **M. levator scapulae inferior** (level 5, Figs. 10, 15). This muscle originates along a gently curved, circular arc, from the entire width of the exoccipital bone along its ventral margin, excluding the medialmost portion between the two occipital condyles. The muscle runs posterolaterally underneath the suprascapula, finally inserting on the ventral surface of this cartilage, near its posterior margin, lateral to the m. serratus medius insertion. The muscle acts to protract the shoulder girdle.
11. **M. serratus superior** (level 5, Figs. 10, 15). This long, slender muscle originates from the transverse process of the fourth vertebra. From here it runs anteromedially, along the lateral margin of the m. rhomboideus posterior (with which it may sometimes be partially fused), to insert alongside the latter muscle on the deep or ventral surface of the suprascapula, near its medial margin, and just posterior to the insertion of the m. rhomboideus anterior. The muscle acts to retract and abduct the shoulder girdle, and to draw it closer to the vertebral column.
12. **M. serratus medius** (level 5, Figs. 10, 15). This short but stout muscle is completely hidden beneath the suprascapula. It originates from the transverse process of the third vertebra, and also to some extent from a tendinous arch connecting this with the transverse process of the fourth vertebra. From here it rises dorsally, though inclined anteromedially, to insert on the ventral surface of the suprascapula, along an arc which lies just lateral to the insertions of the serratus superior, rhomboideus anterior and posterior, and levator scapulae anterior muscles. The muscle acts to depress the shoulder girdle, and to draw it closer to the vertebral column.
13. **M. serratus inferior** (level 5, Figs. 10, 15). This long muscle originates in two portions, from the transverse processes of vertebrae 3 and 4. Both portions then run laterally and slightly anteriorly, as they converge

with each other to insert on the medial or inner surface of the scapula (not the suprascapula), along its posterior margin, somewhat ventral to the insertion of the m. levator scapulae inferior. The muscle acts to adduct the shoulder girdle, and to draw it closer to the vertebral column.

The following four muscles (Nos. 14–17) support the ventral abdominal wall, holding in the viscera and preventing their sagging out from the abdominal cavity. All four are extremely broad and sheetlike, and can be most easily distinguished by the direction of their fibers.

14. **M. cutaneus abdominis** (level 3, Fig. 10). This narrow muscle originates from the cartilaginous pubis, deeply buried in the pelvic region. From here, the fibers pass laterally and then dorsally diverging as they reach around the lateral side of the m. tensor fasciae latae, and finally achieving insertion to the dermis of the skin in the region overlying the ilium and the space between ilium and urostyle. Although this muscle functions as a tensor of the skin in this region, its innervation shows it to be closely related to the next three muscles, which support the abdominal wall.
15. **M. rectus abdominis** (level 1, Figs. 11, 12). This flat muscle is the most superficial of the muscles of the abdominal wall; its fibers run longitudinally, paralleling the ventral midline on either side. The origin is from the pubic symphysis by means of a strong but narrow tendon, in common with the abdominalis portion of the pectoralis muscle. As the muscle passes forward from here, it is subdivided several times by a series of *transverse*, *tendinous inscriptions*, and also by a white line, the **linea alba**, which runs along the ventral midline to separate the right and left rectus abdominis muscles. The muscle continues anteriorly beneath the sternum, where some fibers insert on the xiphisternum. The majority of fibers attach to one last tendinous inscription, after which they continue as the sternohyoideus muscle (*q.v.*). The rectus abdominis muscle shortens the belly, supports the abdominal organs, and draws the sternum posteriorly.
16. **M. obliquus externus** (level 1, Figs. 10, 11, 13). This thin sheet of muscle originates from the ilium and from the fascia which covers the epaxial muscles. From here, its fibers pass obliquely caudally and ventrally, to insert on the aponeurosis which immediately underlies the m. rectus abdominis. The external oblique acts to support the abdominal organs, and to draw the abdominal wall anteriorly while stretching it transversely.
17. **M. obliquus internus** (level 1, Figs. 10, 11, 13). This muscle, also known as the m. transversus abdominis, is a thin sheet of muscle, closely applied to the previous muscle, but differing in the direction of

its fibers. From an origin along the ilium, the dorsal fascia (of the epaxial muscles) and the transverse processes of the vertebrae, the fibers pass obliquely anteriorly and ventrally, beneath the abdominalis portion of the pectoralis muscle, to insert on the anterior portion of the same aponeurosis to which the external oblique attaches.

18. **M. sphincter ani** (level 2, Fig. 10). The fibers of this muscle encircle the anus. Some of the fibers diverge tangentially from the rest to run anteriorly and insert upon the posterior tip of the urostyle. This muscle functions as an anal sphincter.

In addition to the muscles listed above, certain other muscles lie in the trunk region. Of these, the m. cucullaris is treated under the muscles of the head (from which it is derived), while three other muscles (m. latissimus dorsi, m. pectoralis, m. cutaneus pectoris) are derived from, and listed among, the muscles of the pectoral girdle.

G. MUSCLES OF THE HEAD REGION

The muscles of the head are as diverse in derivation as they are in function. The muscles of the eye, and several of the hyoid muscles, are derived from somites (blocks of mesoderm), while the remaining muscles are derived from the fleshy investment of the gill arches (see Section J, below, for further treatment). The head contains six muscles associated with the movements of the eyeball, five with the opening and closing of the jaws, and an even dozen with the movements of the tongue, the floor of the mouth, and the hyoid apparatus. Two remaining muscles, the m. cucullaris and m. interscapularis, though derived from muscles of the head, have become associated with the movements of the shoulder girdle.

Muscles of the Eyeball

The following six muscles are concerned with the movements of the eyeball. The four rectus muscles originate together in the posteromedial corner of the orbit, close to the optic nerve; the superior and inferior oblique muscles originate in the anteromedial corner of the orbit. All six muscles attach to the sclera of the eyeball.

19. **M. rectus superior** (level 3, Fig. 15). Deflects the eyeball dorsally.
20. **M. rectus medialis** (level 3, Fig. 15); also called rectus internus. Deflects the eyeball medially.
21. **M. rectus inferior** (level 3, Fig. 15). Deflects the eyeball ventrally.
22. **M. obliquus inferior** (level 3, Fig. 15). Runs ventrolaterally, to insert low on the eyeball. Rotates the eyeball laterally.

23. **M. obliquus superior** (level 3, Fig. 15). Runs dorsolaterally, to insert high on the eyeball. Rotates the eyeball medially.
24. **M. rectus lateralis** (level 3, Fig. 15); also called rectus externus. Deflects the eyeball laterally.

The following five muscles are associated with the movements of the jaws. The first four muscles act to close the jaws, or to elevate the mandible. The m. depressor mandibulae, as its name implies, acts to open the jaws, or to depress the mandible.

25. **M. temporalis** (level 1, Figs. 10, 13, 15). This strong muscle occupies much of the space within the orbit immediately posterior to the eyeball. The majority of the fibers originate from the prootic and the adjacent part of the exoccipital bones, just medial to the tympanic cartilage; a few fibers also originate directly from this cartilage. The muscle passes deep to the quadratojugal and pterygoid bones, to insert on the coronoid process of the mandible. This muscle closes and retracts the lower jaw.
26. **M. masseter major** (level 2, Figs. 13, 15). This muscle originates from the anteroventral quadrant of the tympanic cartilage along its deep or medial surface, and also from the adjacent anteriorly projecting *zygomatic process* of the squamosal bone. Its fibers pass superficial to the pterygoid and quadratojugal bones, to insert on the outer or lateral surface of the posterior portion of the mandible, including the coronoid process. This muscle closes and protracts the lower jaw.
27. **M. masseter minor** (level 3, Figs. 13, 15). This small, triangular muscle originates from the squamosal (anteroventral margin of the posteroventrally projecting flange) and quadratojugal (lateral surface) bones near the jaw articulation. From their rather localized origin, the fibers diverge, to insert on the lateral surface of the mandible, just beside and posterior to the insertion of the m. masseter major. The anterior portion of the m. masseter minor is covered by m. masseter major; the mandibular branch of the fifth or trigeminal nerve may readily be found separating the two muscles. The m. masseter minor closes the jaws, and also adducts the posterior end of the mandible laterally.
28. **M. pterygoideus** (level 3, Fig. 15). This muscle lies between the m. temporalis and the eyeball. It originates from the prootic and frontoparietal bones, assuming a laterally compressed, flattened shape. The fibers converge posteroventrolaterally, giving way to a long, thin tendon which passes over the pterygoid and quadratojugal bones to insert upon the angulosplenial bone of the mandible, just in front of the jaw articulation, and just posterior to the m. temporalis insertion.

This muscle is a levator, protractor, and contralateral rotator of the lower jaw.

29. **M. depressor mandibulae** (level 1, Figs. 10, 13, 15). This large, triangular muscle has an extensive origin from the fascia overlying the epaxial muscles and the m. temporalis, from the posterodorsally projecting flange of the squamosal, and from the deep or medial surface of the posterior portion of the tympanic cartilage. From here, the fibers converge, passing behind the tympanum at the posterior limit of the skull, and partially covering the scapula. The converging fibers insert onto the extreme posterior end of the lower jaw, behind the jaw articulation, where they serve to depress the lower jaw and open the mouth.

Muscles of the Tongue, the Floor of the Mouth, and the Hyoid Apparatus

30. **M. mylohyoideus** (level 2, Figs. 11, 12, 23). This muscle, also known as the m. submaxillaris, is a broad, flat muscle which spans the floor of the mouth between the two lower jaws. Its origin is from the medial margin of the lower jaw along the angulosplenial bone; the insertion is to a tendinous **raphe** (pronounced "ra-fay" or "ra-fee"), similar to the linea alba, which lies in the ventral midline. This muscle acts to raise the floor of the mouth, and thus to aid in breathing. Cut this muscle to one side of the tendinous raphe, in order to expose the underlying muscles.
31. **M. subhyoideus** (level 3, Figs. 11, 12, 23). This muscle arises from the long anterior cornu of the hyoid apparatus. From here, it passes first ventrally and then laterally, where it comes to lie in parallel with, and just posterior to, the fibers of the m. mylohyoideus; the distinction between these two muscles may not at first be apparent. The insertion is to the same tendinous raphe to which the m. mylohyoideus attaches.
32. **M. submentalis** (level 2, Fig. 12). This small muscle lies deep to the m. mylohyoideus. It consists of a small number of transversely oriented fibers, spanning the mandibular symphysis from side to side, between the mentomeckelian bones. The muscle strengthens the mandibular symphysis.
33. **M. genioglossus** (level 4, Fig. 12). This small muscle lies adjacent to the previous one, but its fibers are longitudinally oriented. The fibers originate from the mandibular symphysis, from which they pass posteriorly, then recurving dorsally and then anteriorly, to enter the substance of the tongue and continue up to its anterior tip. The muscle retracts the tip of the tongue.
34. **M. hyoglossus** (level 3, Fig. 12). This narrow, unpaired muscle lies in the ventral midline on the floor of the mouth, deep to and between the

right and left geniohyoids. From their origin on the body (corpus) of the hyoid cartilage, the fibers run anteriorly, turning dorsally to enter the base of the tongue, and then spreading out to ramify in its substance. The muscle acts to achieve various movements of the tongue, including the rapid protraction and eversion so useful in the catching of insects.

35. **M. geniohyoideus** (level 2, Fig. 12). This paired muscle lies to either side of the ventral midline, covering the previous muscle, and under the cover of the m. mylohyoideus. The fibers originate from the mandible, in the region of the mandibular symphysis; from here they pass directly posteriorly, finally dividing into a larger medial portion, which inserts onto the corpus of the hyoid cartilage, and a smaller lateral portion, which inserts on the posterior cornu. The muscle pulls the hyoid forward.
36. **M. sternohyoideus** (level 2, Fig. 12). This elongated muscle originates from the sternum, from the adjacent portion of the coracoid bone, and from the m. rectus abdominis with the intervention of a tendinous inscription. Passing deep to the coracoid and clavicle, this muscle runs anteriorly, to insert on the corpus of the hyoid cartilage, where its fibers intermingle with those of the m. geniohyoideus. It draws the hyoid posteriorly.
37. **M. omohyoideus** (level 3, Fig. 12). This muscle originates from the inner side of the scapula near its posterior border, and runs anteromedially to insert on the ventral surface of the corpus of the hyoid cartilage. Portions of this muscle lie under the cover of the m. sternohyoideus and m. geniohyoideus. The muscle draws the scapula anteromedially, and the hyoid cartilage posteriorly and laterally.
38. **M. petrohyoideus anterior** (level 4, Figs. 12, 23, 33). This flat, triangular muscle originates from a restricted area on the prootic bone of the skull, adjacent to the tympanic cartilage. From here, it broadens out as it wraps around ventrally and then medially, to insert laterally on the ventral surface of the hyoid corpus. The muscle functions as a constrictor of the pharynx.

39.–41. **Mm. petrohyoidei posteriores I, II, and III** (level 4, Figs. 12, 23, 33). These three long, thin muscles originate together, just posterior to the m. petrohyoideus anterior, in the region just above and behind the ear, largely from the prootic bone of the skull. From here, they pass ventrally and diverge, inserting separately on the posterior or thyroid cornu of the hyoid apparatus: m. petrohyoideus posterior I near the base of the cornu, m. petrohyoideus posterior II in the middle of the cornu, and m. petrohyoideus posterior III near its tip. All three muscles are important in the act of swallowing.

The following two muscles are associated with the movements of the shoulder girdle, but their innervation by the vagus nerve (a cranial nerve) betrays the fact that they are derived from the head. The m. cucullaris not only shares a common innervation with the posterior petrohyoid muscles, but its proximal portion is still in sequence with these.

42. **M. cucullaris** (level 3, Figs. 10, 15, 23). This muscle, corresponding to the trapezius and sternocleidomastoideus muscles of mammals, takes origin from the head, in sequence with the posterior petrohyoid muscles, but inserts on the shoulder girdle. It arises from the prootic and exoccipital bones near their junction, under the cover of the m. depressor mandibulae. From here, it passes posteroventrally to insert on the deep or medial surface of the scapula near its anterior border, between the m. interscapularis and the scapular portion of the m. deltoideus. The cucullaris draws the shoulder girdle cranially and medially, and the head posteroventrally, rotating it also contralaterally.
43. **M. interscapularis** (level 5, Fig. 15). This deep, triangular muscle lies on the deep or inner surface of the shoulder girdle. Its broad origin is confined to the ventral half of the suprascapula along its anterior two-thirds, at the level where the m. levator scapulae inferior inserts. From here, its fibers converge medially to a point, crowding their way between the m. deltoideus and the m. coracobrachialis brevis, and inserting onto the same ridge on the deep surface of the scapula from which these latter two muscles originate. The muscle decreases the angle between the scapula and suprascapula.

H. MUSCLES OF THE SHOULDER GIRDLE AND FORELIMB

Listed below are the true appendicular muscles of the shoulder girdle and forelimb (see Section J, below). In addition to these, certain other muscles which lie in the region of the shoulder girdle have already been described: the rhomboid, serratus, and levator scapulae muscles (Nos. 7–13) among the trunk muscles, and the cucullaris and interscapularis (Nos. 42–43) among the muscles of the head, from which they are derived.

Dorsal (Extensor) Muscles of the Shoulder

44. **M. latissimus dorsi** (level 1, Figs, 10, 13, 15). This fan-shaped muscle originates from the deep surface of the dorsal aponeurosis which overlies the epaxial muscles of the back, along a circular arc which runs from the level of the third vertebra posteriorly and then laterally to the level of the fifth or sixth vertebra. From this origin, the muscle fibers converge as they pass anterolaterally, partially covering the posterior

margin of the scapula and its musculature. The fibers then insert upon a flat tendon, which fuses with the tendon of the m. dorsalis scapulae before inserting on the lateral side of a crest, which runs along the ventral side of the humerus. The m. latissimus dorsi is an extensor (or abductor) of the humerus, and also a retractor of the same.

45. **M. dorsalis scapulae** (level 3, Figs. 10, 13, 15). This broad, triangular muscle covers the majority of the shoulder girdle; its anterior border is in turn covered by the m. depressor mandibulae, and its posterior border by the m. latissimus dorsi. Its origin is along a slightly curved line, parallel to the vertebral column, and extending across the anterior two-thirds of the suprascapula, leaving its medialmost portion uncovered. From here, the fibers converge laterally upon a flat tendon, which fuses with the tendon of the m. latissimus dorsi before inserting on the lateral side of a crest running along the ventral side of the humerus. The m. dorsalis scapulae is an extensor (abductor) and a medial rotator of the humerus.
46. **M. deltoideus** (level 2, Figs. 10, 11, 13, 14). This muscle, a strong protractor of the humerus, is divisible into three distinct portions:

    Pars episternalis. The episternal portion of the deltoid muscle is long and thin. It runs from the margin of the episternum and omosternum to insert, along with half of the *pars scapularis*, on the ulnar epicondyle at the distal end of the humerus.

    Pars clavicularis. The clavicular portion is very small, and runs from the lateral end of the clavicle to a crest (sometimes called the deltoid ridge) that runs along the ventral side of the humerus, inserting on the proximal half of this crest from the medial or anterior side.

    Pars scapularis. The scapular portion is by far the strongest part of the deltoid muscle. Its origin is from the scapula along its anterior border (especially at the tip), and along the inner or medial surface adjacent thereto. The more ventromedially originating fibers insert, together with the *pars clavicularis*, on the proximal half of the deltoid ridge of the humerus from the medial or anterior side. The more dorsolaterally originating fibers insert, together with the *pars episternalis*, on the ulnar epicondyle at the distal end of the humerus. The *pars scapularis* of the deltoid is an abductor and lateral rotator of the arm as well as a protractor.

Ventral (Flexor) Muscles of the Shoulder Region

47. **M. cutaneus pectoris** (level 2, Fig. 11). This muscle lies superficially, beneath the ventral skin in the pectoral region. The fibers originate from the cartilaginous xiphisternum, and from the anterior margin of the sheath which overlies the rectus abdominis. Between the diverging abdominal portions of the right and left pectoralis muscles, the

fibers of the m. cutaneus pectoris run directly anteriorly, where they insert into the dermis of the skin overlying the medial half of each clavicle. The anterior sternal and posterior sternal portions of the m. pectoralis originate beneath this muscle. The right and left mm. cutaneus pectoris are separated across the ventral midline by a space of only moderate width. The muscle acts to tense the skin which overlies the pectoral girdle.

48. **M. pectoralis** (level 1, Figs. 11, 13, 14). This large muscle occupies the greater part of the pectoral region on the ventral side of the body. It consists of three portions, as follows:

    Pars sternalis anterior. The anterior sternalis portion originates from the omosternum, in contact with its partner of the opposite side of the body, and from the adjacent medial end of the coracoid. It partially hides the posterior half of the m. coracoradialis. Its fibers converge laterally, to insert on the deltoid ridge (ventral bony crest) of the humerus from the medial or posterior side, by means of a short tendon which overlies that of the m. coracoradialis.

    Pars sternalis posterior. The posterior sternalis portion originates from the mesosternum and xiphisternum, somewhat separated from the anterior sternalis portion, but more so from the abdominalis portion. Its fibers converge laterally and insert by means of a short tendon onto the very base of the deltoid ridge, deep to the *pars sternalis anterior* and the m. coracoradialis.

    Pars abdominalis. The abdominal portion, largest of the three, originates from the aponeurosis which covers the rectus abdominalis muscle, and also from the pubic symphysis, by means of fibers which intermingle with those of the m. rectus abdominalis. The fibers pass anteriorly and somewhat laterally as they converge onto a strong tendon, which crosses over that of the m. coracoradialis to insert on the deltoid ridge of the humerus from the medial or posterior side, somewhat distal to the insertion of the other two portions.

    All three parts of the m. pectoralis act to adduct and retract the humerus, and also to rotate it medially.

49. **M. coracoradialis** (level 2, Figs. 11, 14). This rather strong muscle originates from the episternum, the omosternum, and the medial portions of the clavicle and coracoid. Its posterior half is covered by the anterior sternalis portion of the m. pectoralis. The fibers converge laterally onto a long, strong tendon, which begins near the shoulder joint. This tendon then passes between the abdominalis portion and the other two portions of the m. pectoralis, along the deltoid ridge, and between the two separately inserting halves of the *pars scapularis* of the m. deltoideus, to insert just beyond the elbow joint on the proximal end of the radioulna.

50. **M. coracobrachialis longus** (level 3, Fig. 14). This long but thin

muscle lies along the coracoid, under cover of the *pars sternalis posterior* of the m. pectoralis. From here it runs laterally, beneath all parts of the m. pectoralis, to insert about halfway down the humerus along its medial or posterior side, deep and somewhat distal to the m. coracobrachialis brevis insertion. This muscle adducts and retracts the arm.

51. **M. coracobrachialis brevis** (level 3, Fig. 14). This muscle originates, partially in common with the previous muscle, from the lateral half of the coracoid along its posterior border, and from the adjacent part of the scapula near its posterior margin. The fibers converge strongly toward the head of the humerus, to insert most proximally on the posterior or medial side of the humerus, including a small tubercle just adjacent to the head, a fine crest passing distally therefrom, the proximal portion of the deltoid ridge near its base, and the space intervening between these. The muscle is a retractor and a lateral rotator of the arm.

Muscles of the Arm

52. **M. triceps brachii** (level 1, Figs. 10, 13, 15, 16). This muscle, which covers the extensor surface of the humerus, originates in three separate portions or heads. The *long head* originates along the posterior border of the scapula, and also from the glenoid capsule; it passes along the dorsal surface of the humerus. The *medial head* originates along the proximal half of the humerus on its medial side; the *lateral head* originates along the proximal half of the humerus on its anterior or lateral side. All three heads unite distally, finally giving way to a strong tendon, which passes over the elbow to insert on the extreme proximal end of the radioulna. Within this tendon, as it passes over the elbow, a sesamoid cartilage is developed at the point of friction. This cartilage is called the **olecranon**, or *patella ulnaris*; it sometimes becomes fused to (i.e. part of) the ulna. The triceps is the principal extensor of the elbow.
53. **M. anconeus** (level 3, Fig. 15). This muscle, also called *subanconeus*, originates from the distal end of the humerus along its dorsal side, between and deep to the various heads of the triceps. It inserts, in common with the tendon of the m. triceps brachii, onto the olecranon cartilage (ulnar patella).

Flexor Muscles of the Forearm

54. **M. flexor carpi radialis** (level 4, Figs. 11, 13, 14, 16, 17). This muscle is more well developed in males than in females, since it is used both to flex the elbow and to draw the hand over to the radial (thumb) side, actions used by the male in the grasping of the female during mating.

The muscle originates from the distal third of the humerus along its posterior margin; a bony crest is developed in males only to mark this origin. The muscle inserts tendinously onto the intermedium.

55. **M. flexor carpi ulnaris** (level 4, Figs. 11, 14, 17). This muscle originates from the epicondyle on the posterior (medial, ulnar) side of the distal end of the humerus, and runs along the forearm to insert both onto the tendon of the previous muscle, and onto its own short tendon, which attaches onto the radiale. This muscle flexes the hand and draws it radially (toward the thumb side).
56. **M. parmaris longus** (level 4, Figs. 11, 14, 17). This muscle originates together with the previous, from the medial epicondyle of the humerus. It passes along the flexor surface of the hand and finally becomes tendinous. The tendon spreads out over the palm of the hand to form the **palmar aponeurosis**, which sends tendinous slips to each of the digits, and also gives rise to the lumbricalis muscles of the fingers. The palmaris longus flexes not only the wrist joint, but also the individual fingers as well.
57. **M. epitrochleocubitalis** (level 4, Fig. 17). This muscle lies in the middle of the flexor surface of the forearm, beside the previous muscle. It originates tendinously from the medial epicondyle of the humerus. Its fibers run diagonally across the radioulna, to insert along the entire length of that bone along its ulnar margin. Fibers of the m. epicondylocubitalis insert along this same line from the opposite side. The muscle flexes the elbow and weakly rotates it laterally.
58. **M. flexor antibrachii medialis** (level 5, Fig. 17). This muscle lies on the flexor side of the forearm, covered by the last three muscles. The muscle originates by means of a tendon from the medial epicondyle of the humerus and also in small part from the deep part of the m. flexor carpi radialis. The fibers insert directly onto the radioulna, along its flexor surface, and somewhat to the radial side. The shortest fibers insert just distal to the insertion of the m. coracoradialis; the longest fibers continue to the distal end of the radioulna. The muscle is a flexor of the elbow.
59. **M. ulnocarpalis** (level 5, Fig. 17). This deep muscle arises under the cover of the m. palmaris longus from the distal third of the radioulna along its flexor surface, somewhat to the ulnar side. It inserts tendinously onto the ulnare. It flexes the wrist.

Extensor Muscles of the Forearm

60. **M. flexor antibrachii lateralis superficialis** (level 4, Fig. 17). This muscle originates from the lateral (anterior, radial) epicondyle of the humerus and also, in common with the m. extensor carpi radialis, from the anterior border of the humerus adjacent thereto. It inserts far

distally on the radial side of the radioulna. It acts as a flexor and lateral rotator.

61. **M. extensor carpi radialis** (level 4, Figs. 10, 11, 13, 14, 16, 17). This muscle lies to one side of the previous, and originates both from the lateral epicondyle of the humerus and from the crest reaching proximally therefrom. The muscle inserts onto a tendon, which becomes part of the joint capsule of the wrist, attached both to the centrale (intermedium) and the distal end of the radioulna. The muscle flexes the elbow, extends the wrist, and draws the wrist radially.
62. **M. abductor indicis longus** (level 4, Figs. 10, 13, 16). This muscle lies on the lateral side of the forearm, where it may be seen in between the m. flexor antibrachii lateralis superficialis and the m. extensor digitorum communis longus. It originates in several portions, from the lateral epicondyle of the humerus, from the lateral margin of the radioulna, and from the radiale. The fibers of all these portions converge, as they cross over the extensor surface of the forearm diagonally, to insert onto the radial (thumb) side of the second metacarpal near its base. The muscle abducts the second finger.
63. **M. extensor digitorum communis longus** (level 4, Figs. 10, 13, 16). This muscle originates from the lateral epicondyle of the humerus. Over the back of the hand, the muscle divides into three portions, which enter the third, fourth, and fifth fingers to connect with the m. extensor brevis superficialis of each. The muscle extends the hand and the last three fingers.
64. **M. extensor carpi ulnaris** (level 4, Figs. 10, 16). This muscle originates together with the previous one from the lateral epicondyle of the humerus and from the capsule of the elbow joint. Along the forearm, it lies between the previous muscle and the m. epicondylocubitalis. Upon reaching the hand, it divides into two portions, one inserting onto the ulnare, the other onto the fused distal carpals 3–5. It extends the wrist.
65. **M. epicondylocubitalis** (level 4, Fig. 17). This muscle arises by a short tendon from the lateral epicondyle of the humerus. Its fibers insert diagonally onto a line running along the ulnar side of the forearm; the m. epitrochleocubitalis inserts onto this same line from the opposite side. This muscle flexes the elbow and rotates the forearm medially.
66. **M. flexor antibrachii lateralis profundus** (level 5, Fig. 17). This deep muscle lies beneath the m. extensor carpi radialis and the m. flexor antibrachii lateralis superficialis. It arises tendinously from the lateral epicondyle of the humerus and inserts along the radial margin of the radioulna, somewhat toward the flexor side. It rotates the forearm laterally.

Muscles of the Hand

67. **M. palmaris profundus** (level 5, Fig. 17). This muscle originates from the distal end of the radioulna along its ulnar border. From here, the fibers diverge as the muscle becomes tendinous, fusing with the deep surface of the palmar aponeurosis and its derivative tendons. The muscle flexes the palmar aponeurosis (thus, the fingers) and draws it also toward the ulnar side.
68. *Muscles of the individual fingers (ventral or flexor side)* (level 6). There are 27 ventral or flexor muscles belonging to the individual fingers, including 2 for the thumb, 4 for the second digit, 3 for the third, 6 for the fourth, 9 for the fifth, and 3 running across the metacarpals.

Muscle	Group
M. abductor pollicis	Thumb
M. adductor pollicis	
M. flexor indicis superficialis	Digit II
profundus	
M. flexor teres indicis	
M. lumbricalis brevis indicis	
M. opponens indicis	
M. lumbricalis brevis digiti III	Digit III
M. flexor teres digiti III	
M. flexor ossis metacarpi digiti III	
M. lumbricalis longus digiti IV	Digit IV
M. interphalangealis digiti IV	
M. lumbricalis brevis radialis digiti IV	
M. lumbricalis brevis ulnaris digiti IV	
M. flexor teres digiti IV	
M. flexor ossis metacarpi digiti IV	
M. lumbricalis longus digiti V	Digit V
M. interphalangealis digiti V	
M. lumbricalis brevis radialis digiti V	
M. lumbricalis brevis ulnaris digiti V	
M. adductor proprius digiti V	
M. opponens digiti V	
M. abductor primus digiti V	
M. abductor secundus digiti V	
M. flexor teres digiti V	
M. transversus metacarpi primus	Muscles running across the metacarpals, from one digit to another
M. transversus metacarpi secundus	
M. transversus metacarpi tertius	

69. *Muscles of the individual fingers* (dorsal or extensor side) (level 6). There are 16 dorsal or extensor muscles belonging to the individual

fingers, except the thumb; these include 5 for the second digit, 4 each for the third and fourth digits, and 3 for the fifth digit.

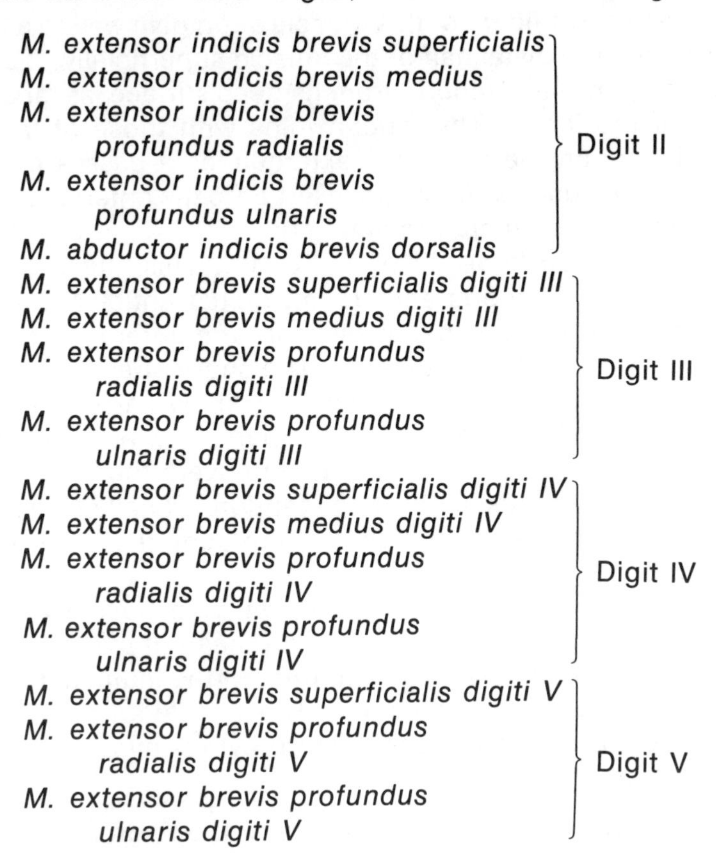

M. extensor indicis brevis superficialis
M. extensor indicis brevis medius
M. extensor indicis brevis profundus radialis
M. extensor indicis brevis profundus ulnaris
M. abductor indicis brevis dorsalis
} Digit II

M. extensor brevis superficialis digiti III
M. extensor brevis medius digiti III
M. extensor brevis profundus radialis digiti III
M. extensor brevis profundus ulnaris digiti III
} Digit III

M. extensor brevis superficialis digiti IV
M. extensor brevis medius digiti IV
M. extensor brevis profundus radialis digiti IV
M. extensor brevis profundus ulnaris digiti IV
} Digit IV

M. extensor brevis superficialis digiti V
M. extensor brevis profundus radialis digiti V
M. extensor brevis profundus ulnaris digiti V
} Digit V

I. MUSCLES OF THE PELVIC GIRDLE AND HIND LIMB

Long Muscles of the Hip and Thigh

The long muscles of the hip and thigh are those muscles which stretch from the pelvic girdle to the tibiofibula, acting across both the hip joint and the knee. The first three of these muscles act as flexors of the hip and extensors of the knee. The remaining long muscles of the hip and thigh act as flexors of the knee: five of these also adduct the thigh, while the other three extend the hip joint.

The following three muscles (Nos. 70–72) act together to simultaneously flex the hip and extend the knee. Since they are interconnected with each other, they are often described as a single muscle, the **M. triceps femoris**.

70. **M. vastus internus** (level 2, Fig. 11). This muscle, also known as the m. cruralis, or the anterior head of the triceps, is the strongest of the

three muscles which make up the triceps. It arises by means of a short tendon, from the capsule of the hip joint along its anteroventral margin. The short fibers of this muscle soon give way to a flat tendon, which covers the distal half of the muscle superficially; the remaining fibers insert onto this tendon from its deep surface, as far as the distal portion of the femur. The tendon joins with those of the m. tensor fasciae latae and the m. vastus externus, as it passes over the knee devoid of any muscular fibers, to insert onto the anterior surface of the tibiofibula at its extreme proximal end.

71. **M. tensor fasciae latae** (level 2, Fig. 10). This muscle, also known as the m. rectus femoris anticus, m. rectus anticus femoris, or the medial head of the triceps, arises from the anteroventral rim of the ilium near the middle of its length. The shortest and most superficial fibers soon give way to an aponeurosis known as the **fascia lata**, to whose deep surface the remaining fibers, and also those of the m. vastus externus, insert. The fascia lata covers the ventral portion of the m. vastus internus, with whose tendon it finally fuses just before passing over the knee joint to insert high upon the anterior surface of the tibiofibula.
72. **M. vastus externus** (level 2, Figs. 10, 18). This muscle, also known as the m. gluteus magnus, or the posterior head of the triceps, arises by a tendon from the posterodorsal corner of the ilium, at the base of its long, anteriorly projecting process. The muscle forms a distinct belly, whose fibers insert, like those of the previous muscle, onto the deep surface of the fascia lata.

The following five muscles are adductors of the thigh and flexors of the knee:

73. **M. sartorius** (level 1, Fig. 11). This long, flat muscle lies on the inner or medial side of the thigh. It arises by a series of barely discernible tendinous strands from the anteroventralmost corner of the pubic symphysis. Its parallel fibers run most superficially along the thigh, giving way to a tendon as they approach the knee. The tendon runs across the knee medially, to insert from the posterior side onto the medial or inner surface of the tibiofibula, near its proximal end. The tendinous fibers, as they spread out to form the insertion, meet with the tendinous insertions of the m. semitendinosus and the m. triceps femoris.
74. **M. adductor longus** (level 3, Figs, 11, 19). This long, flat muscle lies under the cover of the m. sartorius. It originates, deep to the m. sartorius origin, from the anteroventralmost corner of the pubic symphysis, by means of a flat tendon. The muscle then passes over the medial or inner side of the thigh, to insert along the distal half of the femur onto the tendon of the m. adductor magnus.

75. **M. adductor magnus** (level 3, Figs. 11, 18, 19). This rather large muscle lies largely covered by the m. sartorius and the m. gracilis major. It originates in three distinct heads: the dorsal and ventral heads arise from the ventral or posteroventral margin of the pelvic girdle (ischium, also pubis), separated from each other by the ventral head of the m. semitendinosus. The accessory head, smallest of the three, arises from the tendon of origin of the ventral head of the semitendinosus muscle; it inserts itself between the other two heads, whereupon all three fuse, passing as far as the distal end of the femur before giving way to the tendon of insertion. This tendon, to which the previous muscle also attaches, inserts onto the medial side of the femur at its distal end; muscle fibers of the m. adductor magnus also insert directly onto the distal third of the femur, as well as onto the capsule of the knee joint.
76. **M. gracilis major** (level 2, Figs. 11, 19). This strong muscle, also known as the m. rectus internus major, lies along the posteromedial surface of the thigh. It arises by short, tendinous fibers from the posterior margin of the ischium. After crossing the thigh, it gives way to a tendon as it reaches the knee. This tendon then bifurcates, one portion inserting from the medial side onto the proximal end of the tibiofibula, deep to the m. sartorius insertion, while the other portion dives around the tendinous insertion of the m. semitendinosus to insert on the posterior surface of the tibiofibula from the medial side.
77. **M. gracilis minor** (level 2, Figs. 11, 19). This narrow muscle, also known as the m. rectus internus minor, lies beside the previous muscle on the posteromedial margin of the thigh. It originates, along with its partner of the opposite side, from a median raphe which also serves more anteriorly as origin for fibers of the m. rectus abdominis. As it crosses the knee, the muscle inserts to the tendon of the m. gracilis major (*q.v.*).

The following three muscles (Nos. 78–80) extend the hip joint as they flex the knee:

78. **M. ileofibularis** (level 3, Figs. 10, 18, 21). This muscle, also known as the m. biceps femoris, is visible on the dorsal (posterior) side of the thigh, between the m. vastus externus and the m. semimembranosus. It arises tendinously from the ilium, just behind the origin of the m. vastus externus. Over the knee, the muscle passes into a tendon, which then forms a tendinous arch, inserting more proximally onto the medial surface of the distal end of the femur from the posterior side, and more distally, onto the lateral or fibular portion of the tibiofibula, near its proximal end.
79. **M. semimembranosus** (level 2, Figs. 10, 18). This strong muscle begins with a broad origin along the posterodorsal margin of the

ischium. It remains fleshy over the entire length of the thigh, inserting by means of a short tendon onto the posterior surface of the tibiofibula near its proximal end, surrounded by the tendinous arch from which the m. gastrocnemius originates.

80. **M. semitendinosus** (level 3, Figs. 18, 19). This long, thin muscle lies rather dorsally (posteriorly) along the inner or medial side of the thigh. The dorsal head arises by a long, thin tendon from the posterior margin of the ischium, deep to the origin of the m. semimembranosus. The ventral head arises by an even narrower tendon from the ventral part of the ischium, between the dorsal and ventral heads of the m. adductor magnus. The dorsal head of the m. adductor magnus lies between the two heads of the semitendinosus; the accessory head of that muscle arises from the ventral head of the present one. The two heads of the m. semitendinosus fuse over the middle of the thigh, giving rise to a tendon as they approach the knee. The tendon forms a triangular plate as it inserts from the posterior side onto the medial surface of the tibiofibula near its proximal end.

Short Muscles of the Hip and Thigh

The short muscles of the hip and thigh are those muscles which cross the hip joint but do not reach the knee, inserting onto the femur rather than the tibiofibula. At their origins, these muscles lie in three layers: the first four superficially, the next four deep to these, and the m. obturator internus deepest of all.

81. **M. iliacus internus** (level 3, Fig. 18). This muscle arises from the concave margin of the ilium near its base, on the anterior side of the pelvic girdle, and also from the anterior margin of the pelvic girdle on the inner or medial surface of the ilium at its base. It dives between the m. vastus internus and the m. tensor fasciae latae, to insert beside the m. ileofemoralis on the lateral surface of the femur. It acts as an abductor of the femur.
82. **M. iliacus externus** (level 4, Fig. 10, 18). This muscle, also known as the m. gluteus, arises rather sheetlike from the dorsal margin of the ilium along the middle third of its length on the lateral side. The muscle passes between the m. vastus externus and the m. tensor fasciae latae as its fibers converge onto a strong tendon. The tendon inserts rather high on the femur (to its trochanter), between the m. iliacus internus and m. ileofemoralis. It rotates the femur medially.
83. **M. ileofemoralis** (level 4, Fig. 18). This small muscle originates from the tendon of the m. ileofibularis. It inserts onto the lateral surface of the shaft of the femur, between the m. iliacus internus and the m. pyriformis. It adducts and retracts the femur.

84. **M. pyriformis** (level 3, Figs. 10, 18). This narrow muscle arises from the tip of the urostyle, and inserts between the m. vastus externus and m. ileofibularis on one side, and the m. semimembranosus on the other, onto the proximal portion of a crest which runs along the posterodorsal side of the femur. The muscle retracts and adducts the femur.
85. **M. pectineus** (level 4, Figs. 11, 19). This muscle originates from the anteroventral quadrant of the pelvic girdle (including the pubis and the adjacent part of the ilium), deep to the m. sartorius. From here, it runs laterally, to insert beside the m. pyriformis, onto the same long crest, as far distally as the halfway point along the femur. The muscle acts as a protractor of the femur, and also (with the femur protracted) as a flexor.
86. **M. obturator externus** (level 4, Fig. 18). This muscle arises beside the m. pectineus, only more posteriorly. It parallels the latter muscle throughout its length, to insert beside it medially along the posterior side of the femur. This muscle flexes the femur. This muscle is sometimes described together with the next two, under the name m. adductor brevis.
87. **M. quadratus femoris** (level 4, Fig. 18). Arises from the posterior portion of the pelvic girdle under cover of the dorsal head of the m. adductor magnus. The fibers converge to a restricted area of insertion on the femur, proximal and dorsal to that of the m. obturator externus. Adducts the femur.
88. **M. gemellus** (level 4, Fig. 18). This muscle arises most dorsally from the ischium, under cover of the m. semimembranosus. Its fibers converge to an insertion proximal to that of the previous muscle. It adducts and retracts the femur.
89. **M. obturator internus** (level 5, Fig. 18). This muscle is the deepest of all muscles of the hip. It surrounds the acetabulum from three sides, originating anterior, ventral, and posterior thereto. From here, the fibers converge in spiral fashion onto a tendon, by which the muscle inserts high on the posterodorsal margin of the femur, near its head. The muscle adducts the femur and rotates it laterally.

Muscles of the Lower Leg

The first two of these muscles are extensors of the ankle; the remaining four are flexors of this joint. Several of these muscles act across the knee joint as well, usually performing the opposite motion here as across the ankle.

90. **M. gastrocnemius** (level 1, Figs. 10, 11, 21). This muscle, also known as the m. plantaris longus, is by far the strongest muscle of the hind limb, and is the principal muscle used by the frog in jumping. The

muscle arises from a tendinous arch, which connects the adjacent ends of the femur and tibiofibula across the knee joint. A small number of fibers also arise tendinously from the lateral margin of the fascia lata (see No. 71, above) as it crosses the knee capsule. The muscle then develops a very strong, meaty belly (frogs' legs—the gourmet's delight!), on whose lower half a superficial aponeurosis appears. This aponeurosis gives rise to a long, strong tendon, the **tendon of Achilles**, which crosses the "heel" of the ankle joint and then spreads out along the "sole" of the foot to form a **plantar aponeurosis**, which in turn gives rise to other muscles and also sends tendinous strands into each of the several digits. The gastrocnemius muscle is the principal extensor muscle of the ankle joint. (For advanced students: This movement of the ankle, though descriptively an extension, corresponds more closely to the act of *flexion* in the wrist. For this reason, we may unambiguously designate this motion as **plantar flexion**, and the corresponding opposite motion as **dorsiflexion**.)

91. **M. tibialis posterior** (level 3, Figs. 11, 21). This thin muscle lies beneath the m. gastrocnemius and toward the inner (medial) or tibial side along the posterior surface of the tibiofibula. Its fibers take origin all along this bone as far as its distal end, running diagonally to converge upon a tendon which spans the ankle joint to insert on the proximal end of the tibiale. The muscle is a plantar flexor, and also an adductor (pronator) of the foot.
92. **M. peroneus** (level 2, Figs. 10, 20, 21). This long but strong muscle lies along the outer (lateral) side of the leg, between the m. gastrocnemius and the m. tibialis anticus longus. It arises quite far laterally along the extensor surface of the knee by means of a strong but narrow tendon. The muscle inserts by means of two tendons, which together form a tendinous arch. One end of this arch is anchored to the distal end of the tibiofibula, the other to the lateral corner of the fibulare. The tendon of origin of the m. tarsalis anterior (see below, No. 103) passes under this tendinous arch. The m. peroneus extends the knee, dorsiflexes the foot, and pronates it.
93. **M. tibialis anticus longus** (level 2, Figs. 10, 11, 20). This muscle arises by means of a thin, long tendon which crosses obliquely over the extensor side of the knee from the distal end of the femur. The muscle fibers originate to either side of this tendon, forming two bellies over the anterior side of the lower leg. Each of these two bellies ends in a long, narrow tendon, the more medial of the two crossing over the tendon of the m. tibialis posterior to insert nearby on the tibiale, the other tendon diverging laterally to insert on the fibulare. The muscle acts in the extension of the knee and the dorsal flexion of the ankle.
94. **M. extensor cruris** (level 3, Figs. 11, 20). This muscle, also known as

the m. extensor cruris brevis, lies on the lateral side of the lower leg, adjacent to the m. tibialis anticus longus. Its somewhat long and thin tendon of origin arises from the medial side of the distal end of the femur, and crosses the knee just parallel medially to the tendon of the m. tibialis anticus longus. The muscle inserts directly onto the lateral margin of the tibiofibula, near its distal end. It extends the knee.

95. **M. tibialis anticus brevis** (level 3, Fig. 20). This muscle arises from the anterior surface of the tibiofibula over its middle third, between the m. tibialis anticus longus and the m. extensor cruris brevis. The muscle drifts gradually medially as it passes distally, paralleling the medial portion of the m. tibialis anticus longus. The muscle then gives way to a thin tendon, which is crossed over by that of the medial portion of the m. tibialis anticus longus. It then inserts to the proximal end of the tibiale, near the insertion of the m. tibialis anticus longus, and just medial to the insertion of the m. tibialis posterior. The muscle flexes the foot dorsally, and turns the sole inwardly.

Muscles of the Foot

The muscles of the foot are arranged in two groups: the muscles on the sole of the foot (flexor muscles), and the muscles of the dorsum of the foot (extensor muscles). The flexor muscles will be listed first (Nos. 96–102); the extensor muscles will be treated last (Nos. 103–104).

96. **M. tarsalis posterior** (level 5, Figs. 20, 21). This muscle arises from a strong ligament (the **ligamentum calcanei**) which runs from side to side across the plantar surface of the tarsus at its extreme proximal end. This ligament also contains a sesamoid cartilage toward its medial side; fibers of the m. tarsalis posterior originate from this cartilage as well. The muscle inserts onto the plantar surface of the tibiale over the distal two-thirds of its length. The muscle acts in the plantar flexion of the foot, and in turning the sole of the foot outward.
97. **M. plantaris profundus** (level 5, Fig. 21). This muscle originates from the ligamentum calcanei and its sesamoid cartilage, partially overlying the previous muscle. It spreads out over the sole, to insert onto the deep (dorsal) surface of the plantar aponeurosis. It tenses the palmar aponeurosis and thus indirectly flexes the first three digits of the foot.
98. **M. flexor digitorum brevis superficialis** (level 5, Fig. 21). This muscle originates most laterally from the ligamentum calcanei. At the level of the tarsometatarsal joint, the muscle gives rise rather abruptly to a tendon which then trifurcates, sending tendinous slips to each of digits III, IV, and V. The muscle flexes those digits to which it inserts.
99. **M. transversus plantae proximalis** (level 5, Fig. 26). This muscle arises from the distal end of the fibulare, crossing the sole of the foot

transversely as a broad, flat sheet. Its fibers proceed medially, to insert onto the plantar aponeurosis from the deep or dorsal side. The muscle draws the medial portion of this aponeurosis laterally, assisting the lumbricales of the first three digits especially.

100. **M. transversus plantae distalis** (level 5, Fig. 26). This muscle arises from a sesamoid cartilage that lies within the sole of the foot. Its fibers run medially and somewhat distally, some of them inserting onto the plantar aponeurosis, and others onto a tendon, attached to the plantar aponeurosis, from which the m. lumbricalis longus digiti III and m. lumbricalis brevis digiti IV originate. The muscle assists in the fixation of these latter at their origin.
101. **M. intertarsalis** (level 5, Fig. 26). This strong muscle is the deepest of the muscles of the sole. It arises over the proximal two-thirds of the fibulare, and to a lesser extent from the proximal third of the tibiale. The fibers of this muscle converge onto a strong tendon, which inserts onto the plantar surface of the centrale. The muscle adducts the foot and turns the sole outward.
102. *Muscles of the individual toes (plantar or flexor surface)* (level 6). There are 32 plantar or flexor muscles belonging to the individual toes of the foot. These include 1 for the prehallux, 4 for the hallux (big toe), 3 for the second toe, 5 for the third, 9 for the fourth, 6 for the fifth, and 4 running across the metatarsals.

M. abductor prehallucis	Prehallux
M. lumbricalis brevis hallucis	Hallux (Digit I)
M. flexor teres hallucis	
M. abductor brevis plantaris hallucis	
M. opponens hallucis	
M. lumbricalis brevis digiti II	Digit II
M. flexor teres digiti II	
M. flexor ossis metatarsi digiti II	
M. lumbricalis longus digiti III	Digit III
M. interphalangeus digiti III	
M. lumbricalis brevis digiti III	
M. flexor teres digiti III	
M. flexor ossis metatarsi digiti III	
M. lumbricalis longissimus digiti IV	Digit IV
M. interphalangeus distalis digiti IV	
M. lumbricalis longus digiti IV	
M. interphalangeus proximalis digiti IV	
M. lumbricalis brevis lateralis digiti IV	
M. lumbricalis brevis medialis digiti IV	
M. flexor teres digiti IV	
M. flexor ossis metatarsi digiti IV	
M. abductor proprius digiti IV	

Muscle	Group
M. lumbricalis longus digiti V	Digit V
M. interphalangeus digiti V	
M. lumbricalis brevis lateralis digiti V	
M. lumbricalis brevis medialis digiti V	
M. flexor teres digiti V	
M. abductor brevis plantaris digiti V	
M. transversus metatarsi primus	Muscles running across the metatarsals, from one digit to another
M. transversus metatarsi secundus	
M. transversus metatarsi tertius	
M. transversus metatarsi quartus	

The remaining muscles of the foot (Nos. 103–104) lie on its dorsal surface and are primarily extensors.

103. **M. tarsalis anterior** (level 5, Figs. 11, 20). This fairly strong muscle arises by a brief tendon from the posterolateral margin of the tibiofibula at its distal end, under cover of the m. tibialis anticus longus. From here, it runs onto the dorsal surface of the foot, where it inserts onto the tibiale, along its more distal half. The muscle is a dorsiflexor of the foot; it also supinates the foot, so that the sole faces inwardly.
104. *Muscles of the individual toes (dorsal or extensor surface)* (level 6). There are 22 dorsal or extensor muscles of the individual toes, including 5 for the hallux (big toe), 4 for the second digit, 4 for the third, 5 for the fourth, and 4 for the fifth.

Muscle	Group
M. extensor brevis superficialis hallucis	Hallux (Digit I)
M. extensor brevis medius hallucis	
M. extensor brevis profundus lateralis hallucis	
M. extensor brevis profundus medialis hallucis	
M. abductor brevis dorsalis hallucis	
M. extensor brevis superficialis digiti II	Digit II
M. extensor brevis medius digiti II	
M. extensor brevis profundus lateralis digiti II	
M. extensor brevis profundus medialis digiti II	
M. extensor brevis superficialis digiti III	Digit III
M. extensor brevis medius digiti III	
M. extensor brevis profundus lateralis digiti III	
M. extensor brevis profundus medialis digiti III	

Muscle	Digit
M. extensor longus digiti IV	Digit IV
M. extensor brevis medius lateralis digiti IV	Digit IV
M. extensor brevis medius medialis digiti IV	Digit IV
M. extensor brevis profundus lateralis digiti IV	Digit IV
M. extensor brevis profundus medialis digiti IV	Digit IV
M. extensor brevis superficialis digiti V	Digit V
M. extensor brevis profundus lateralis digiti V	Digit V
M. extensor brevis profundus medialis digiti V	Digit V
M. abductor brevis dorsalis digiti V	Digit V

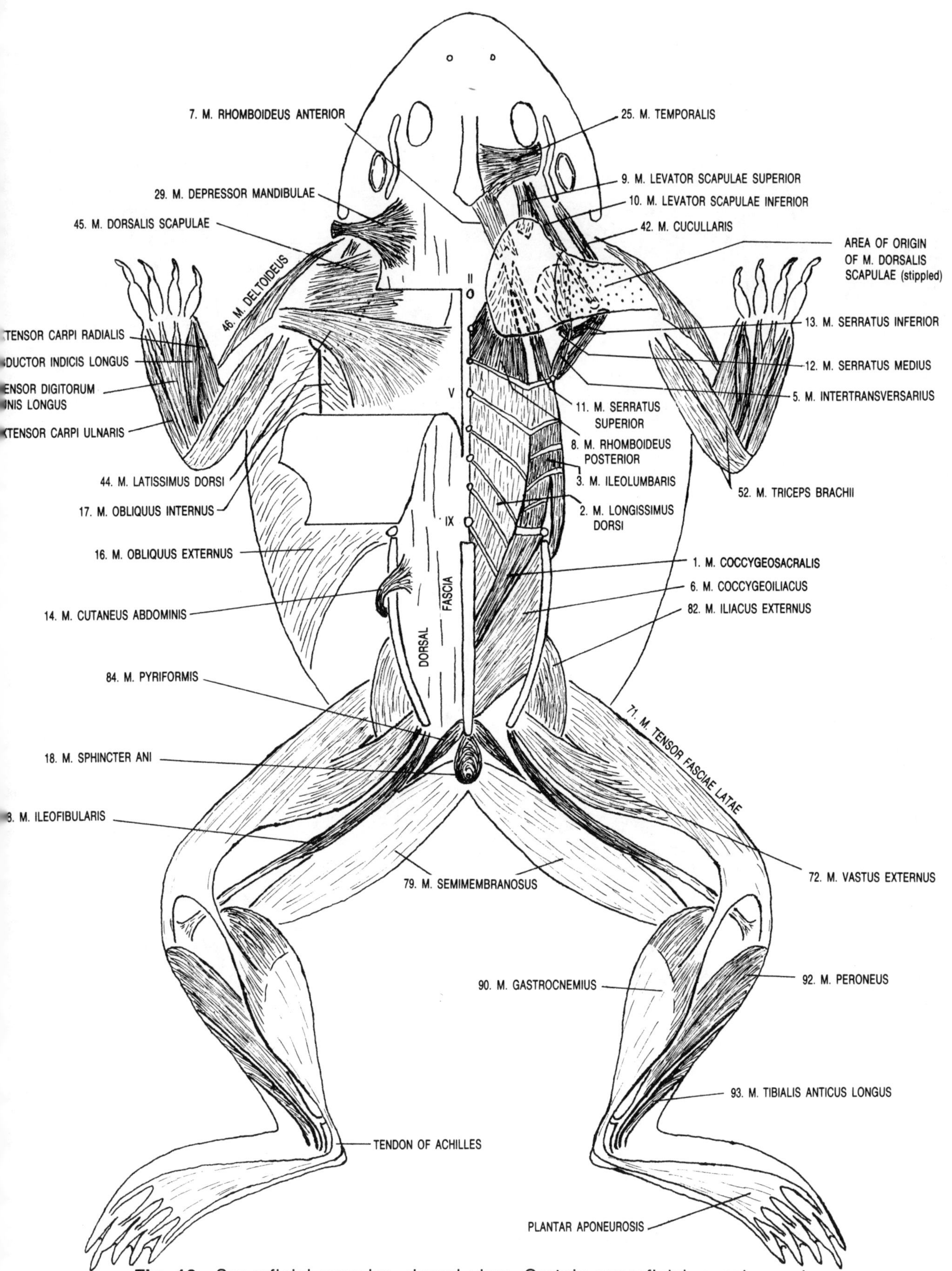

Fig. 10 Superficial muscles, dorsal view. Certain superficial muscles and other overlying structures have been removed on the right side, in order to reveal a slightly deeper layer.

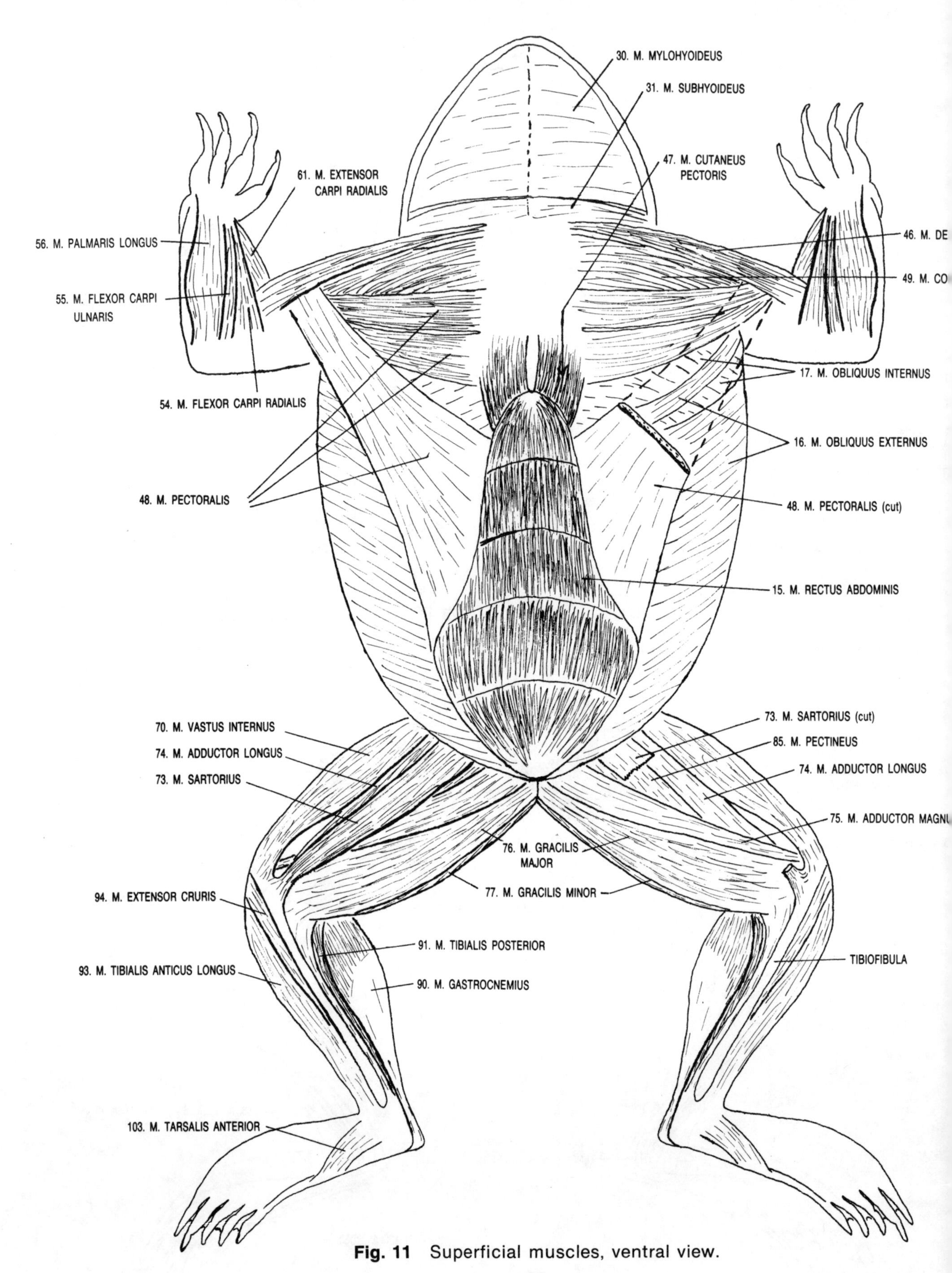

Fig. 11 Superficial muscles, ventral view.

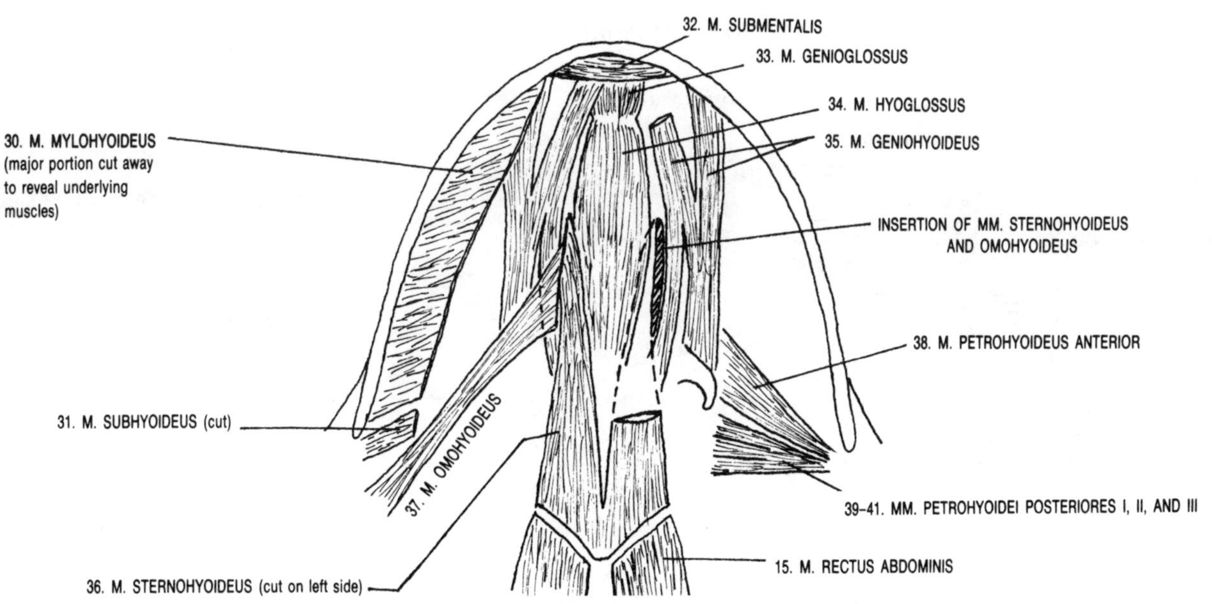

Fig. 12 Muscles of the head region, ventral view. The left side (i.e. the right half of the diagram) represents a somewhat deeper dissection.

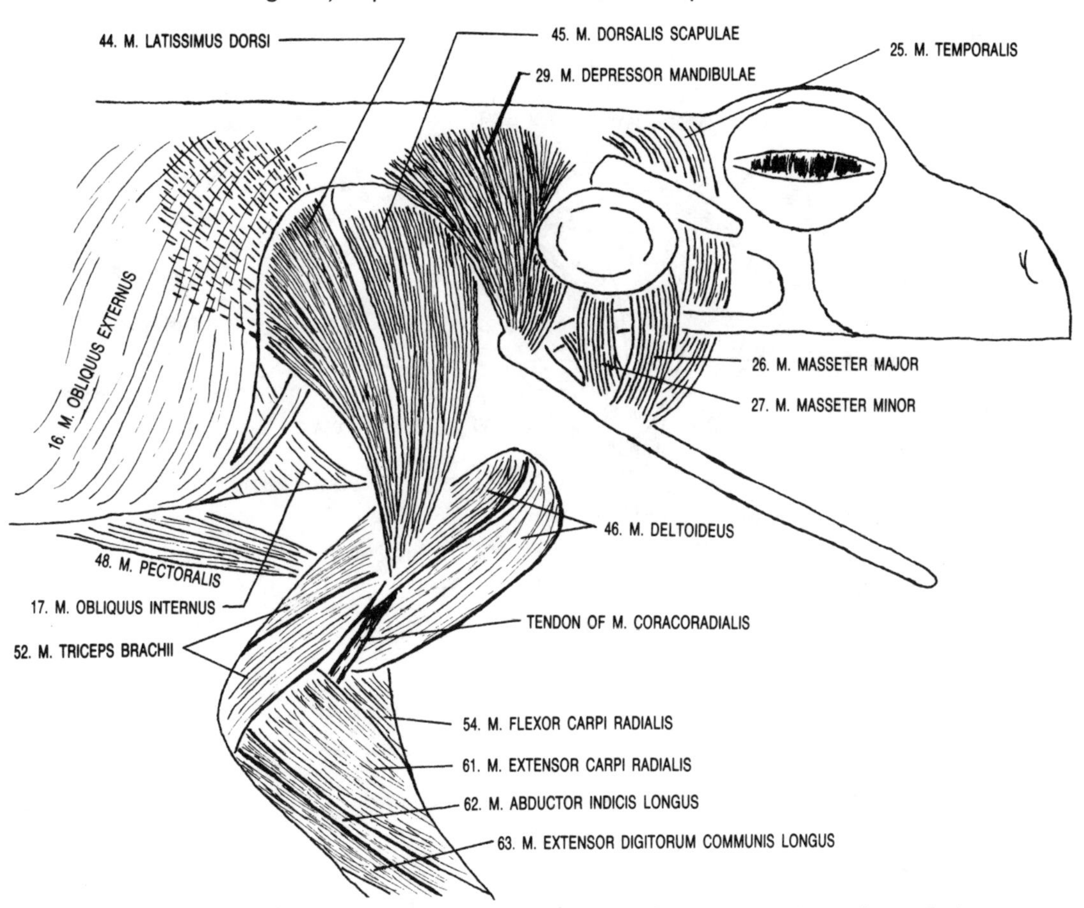

Fig. 13 Superficial muscles of the head and shoulder regions, lateral view.

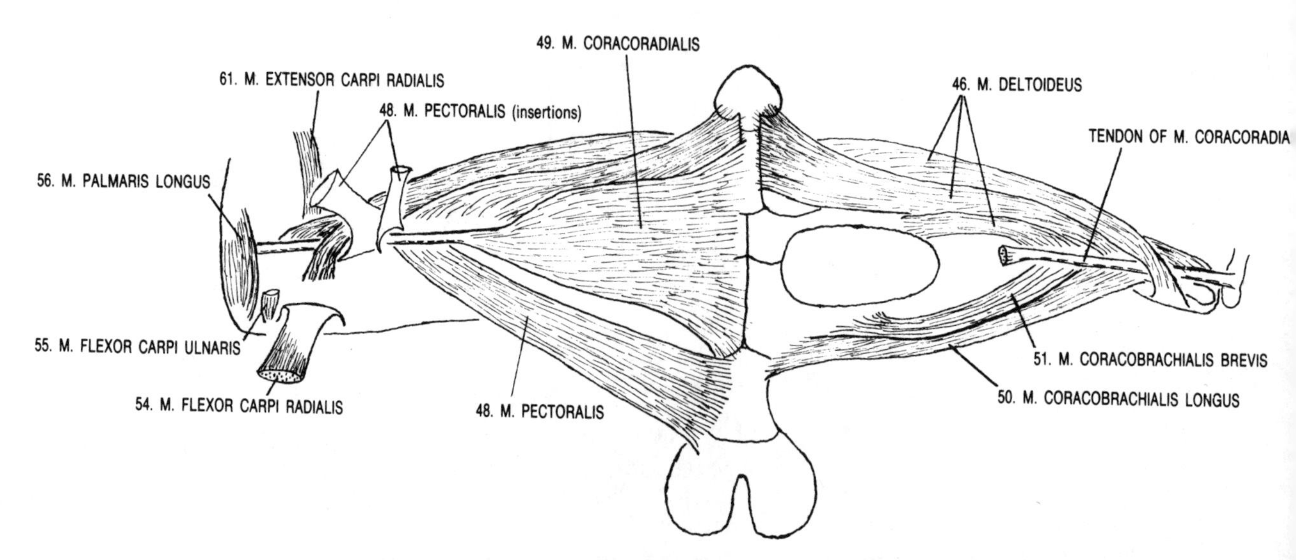

Fig. 14 Pectoral region, deep dissection (ventral view).

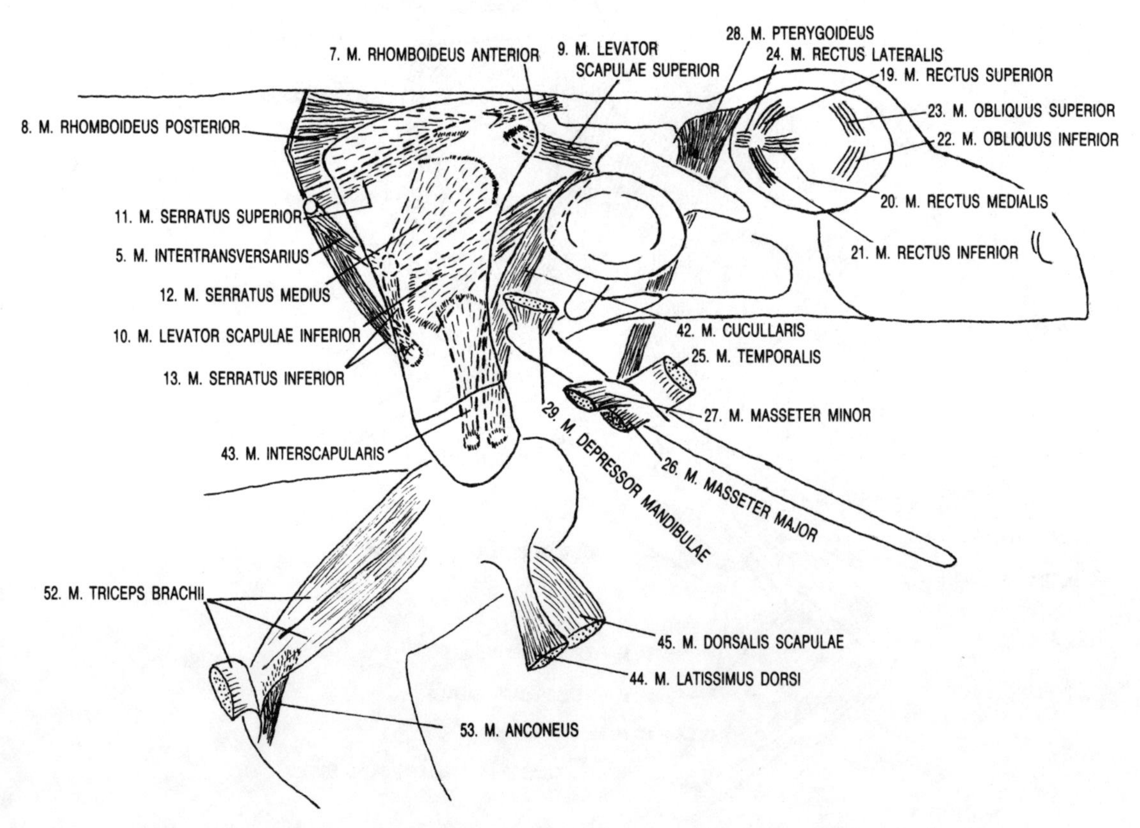

Fig. 15 Head and shoulder regions, deep dissection (lateral view).

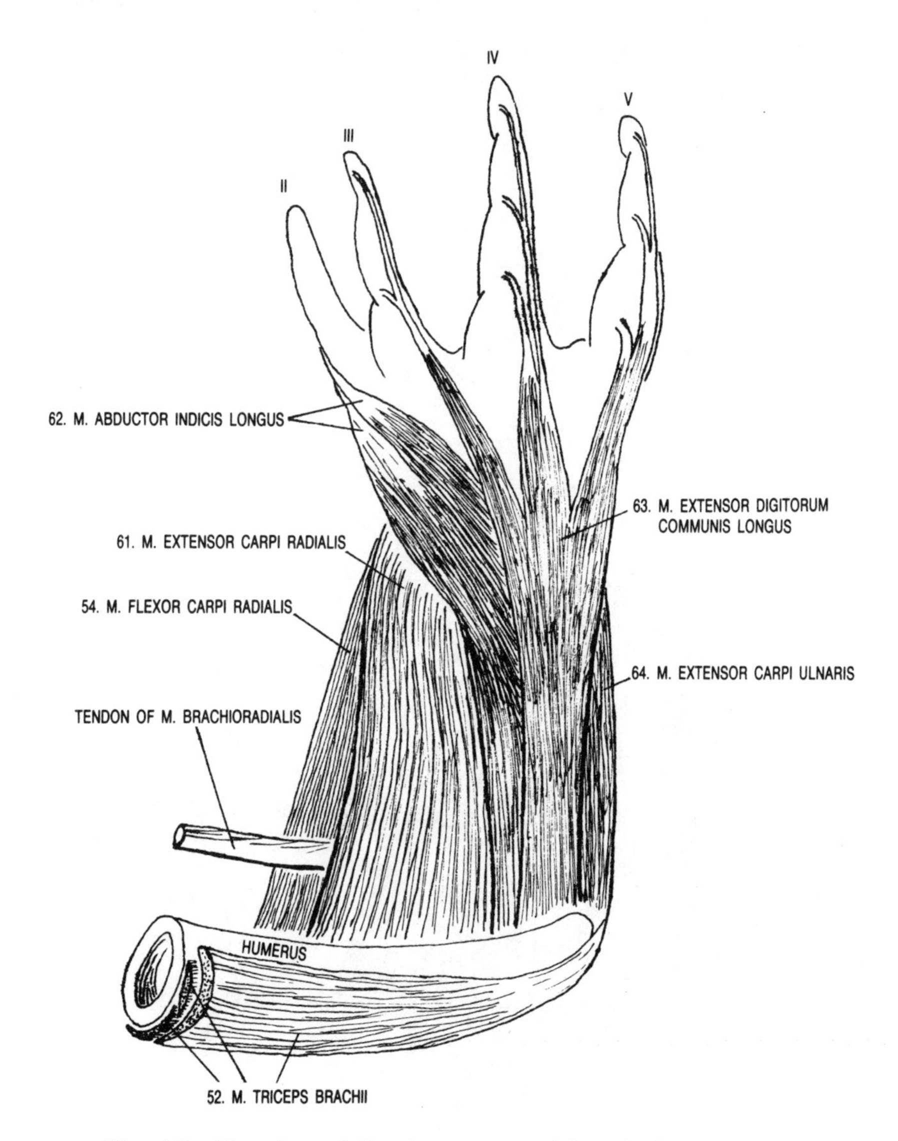

Fig. 16 Muscles of the forearm and hand, dorsal view.

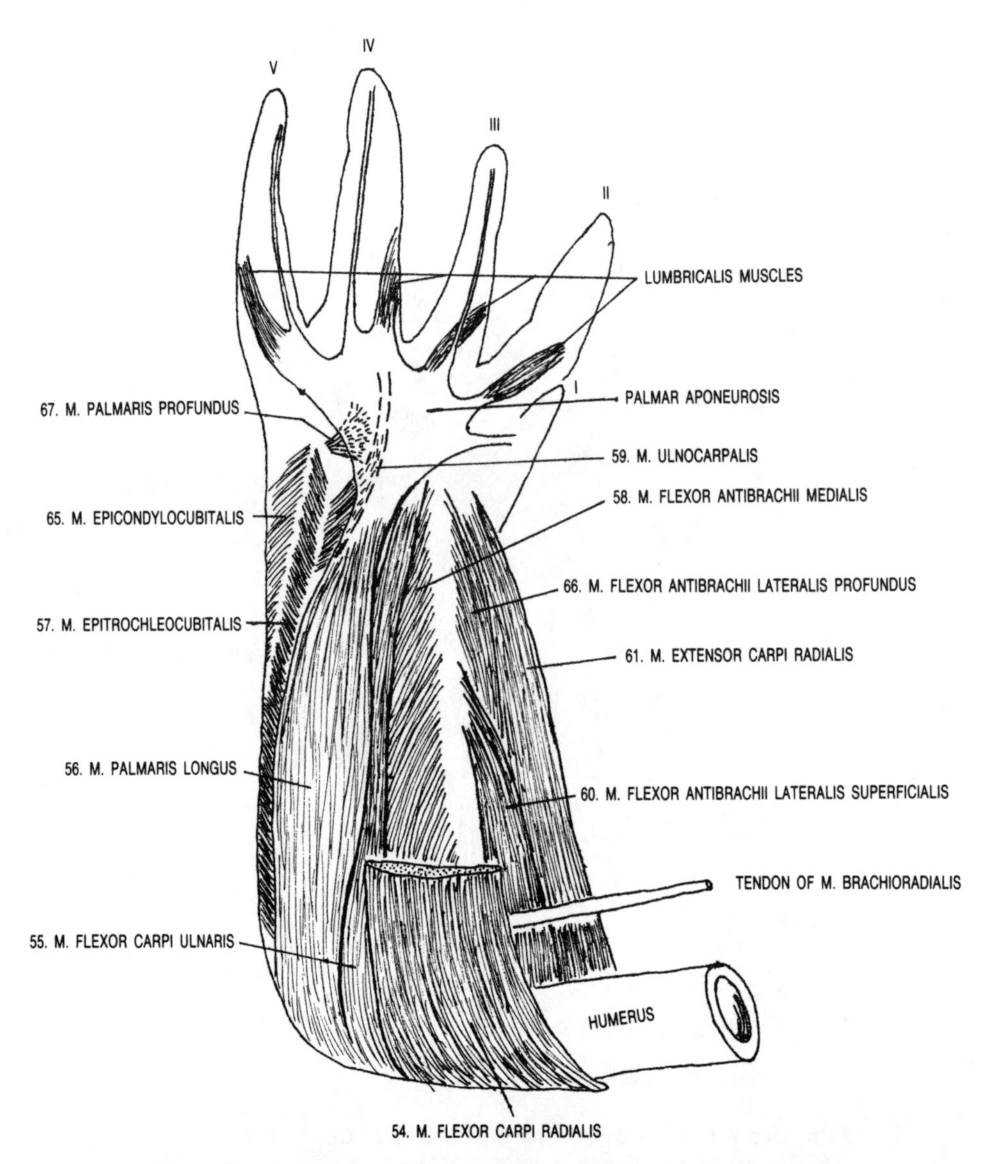

Fig. 17 Muscles of the forearm and hand, ventral view.

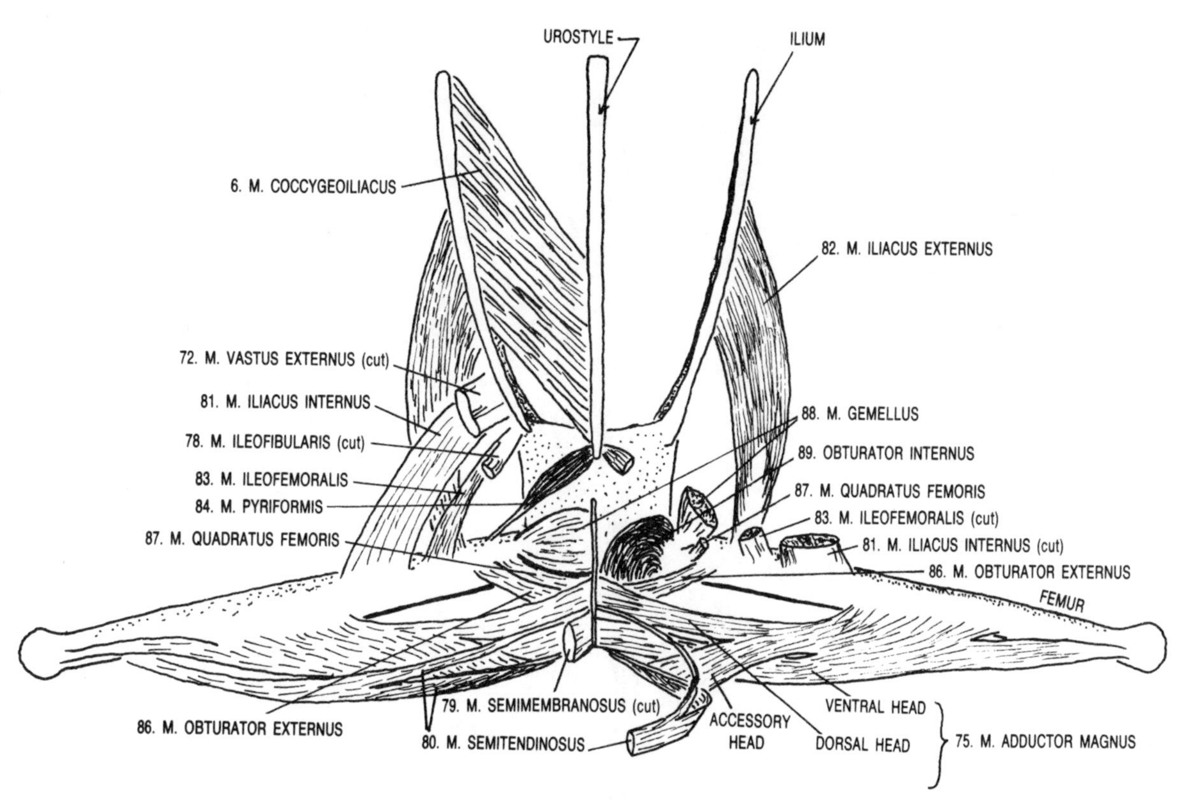

Fig. 18 Deep muscles of the pelvic region, dorsal view.

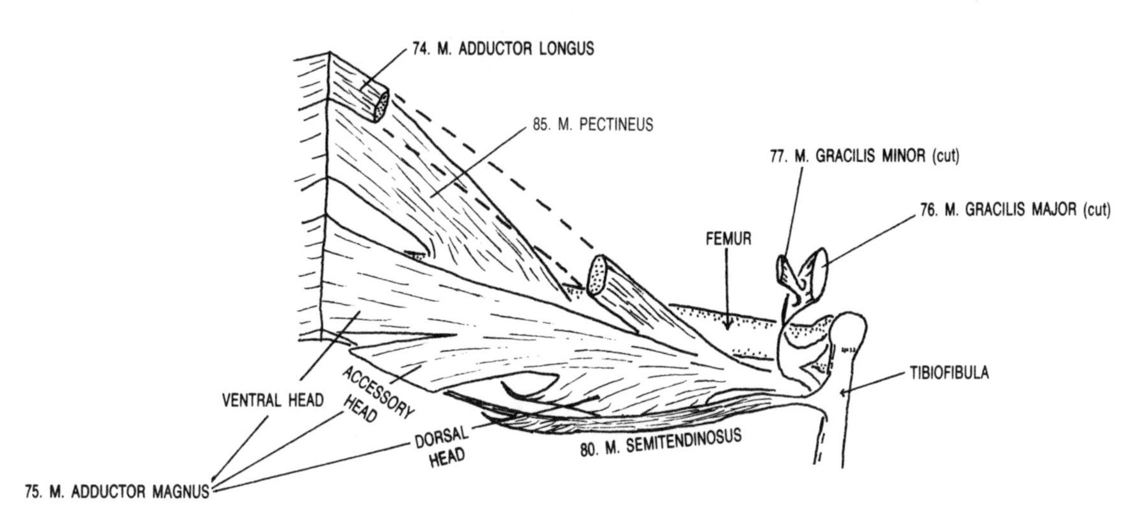

Fig. 19 Deep muscles of the pelvic region, ventral view.

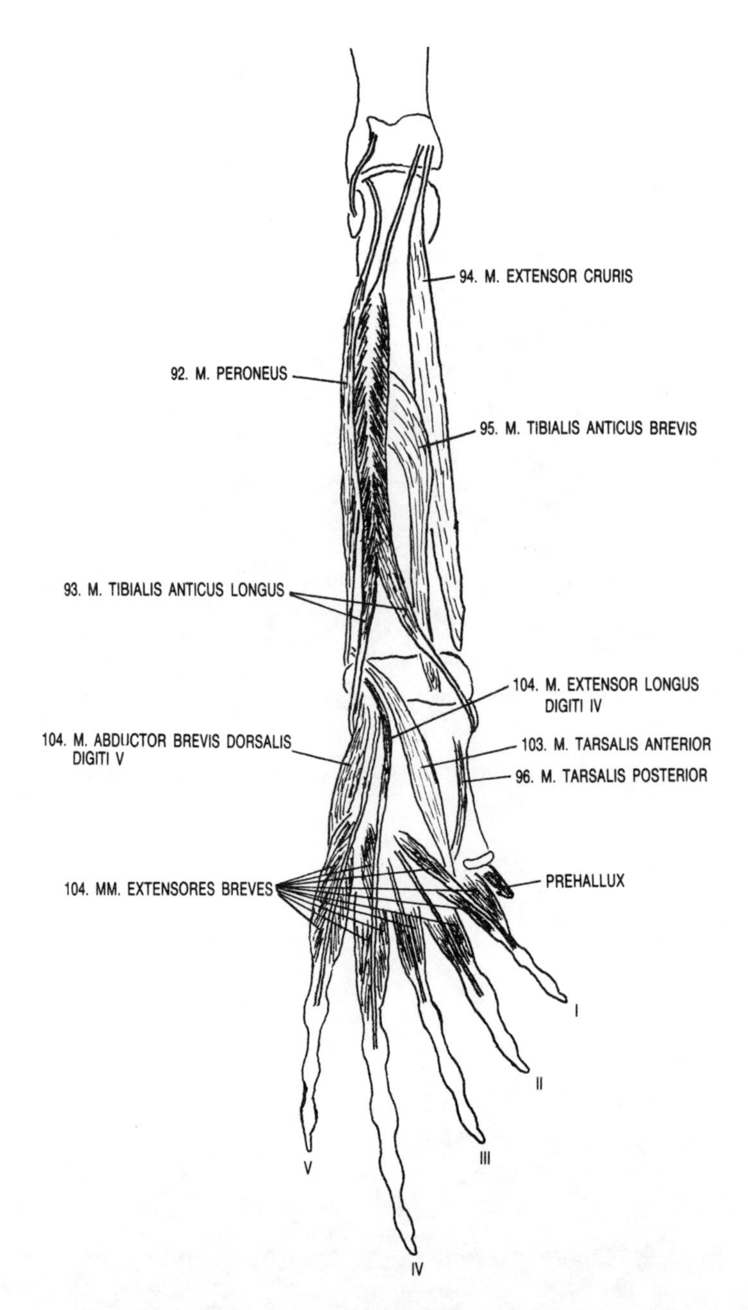

Fig. 20 Muscles of the leg and foot, anterior view.

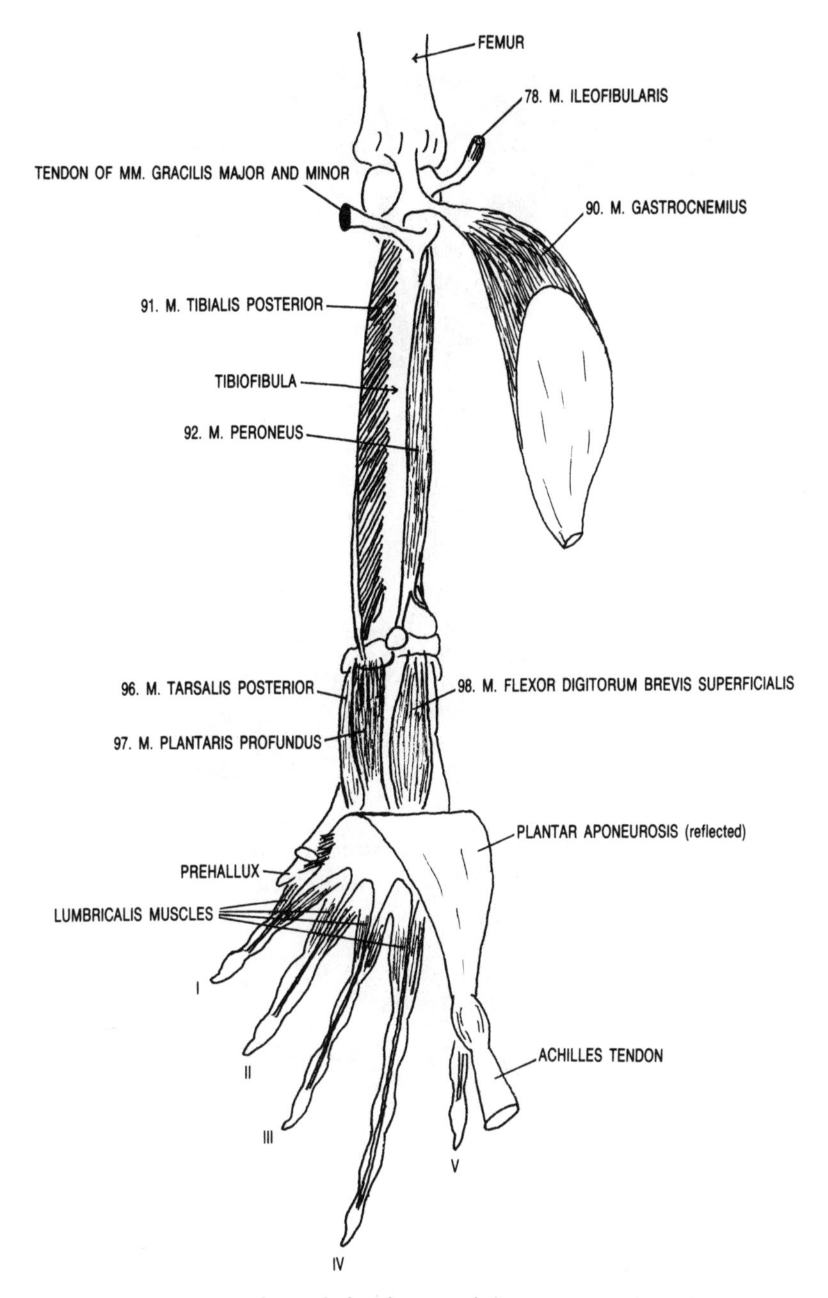

Fig. 21 Muscles of the leg and foot, posterior view.

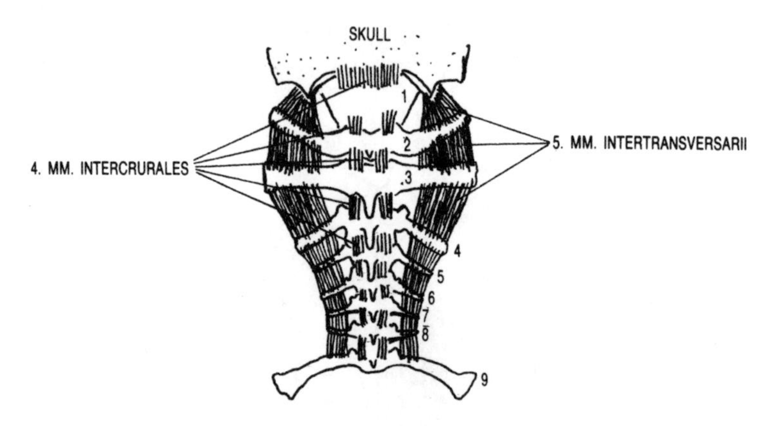

Fig. 22 Mm. intercrurales and intertransversarii, dorsal view.

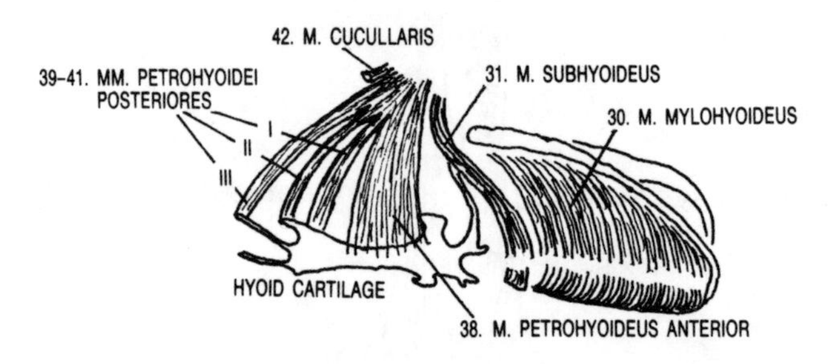

Fig. 23 The petrohyoideus and subhyoideus muscles, and associated structures; oblique ventrolateral view.

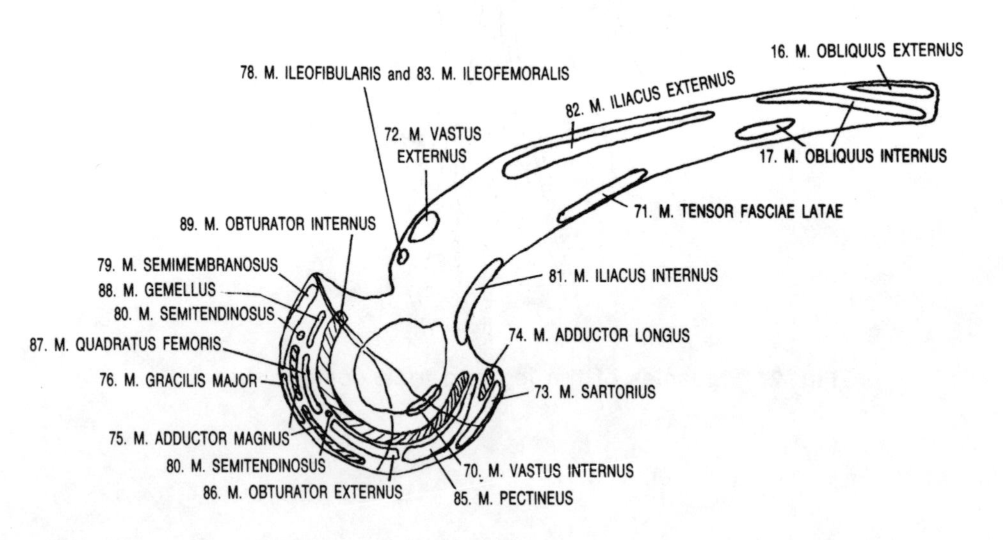

Fig. 24 Lateral view of the pelvic girdle, showing areas of muscle attachment.

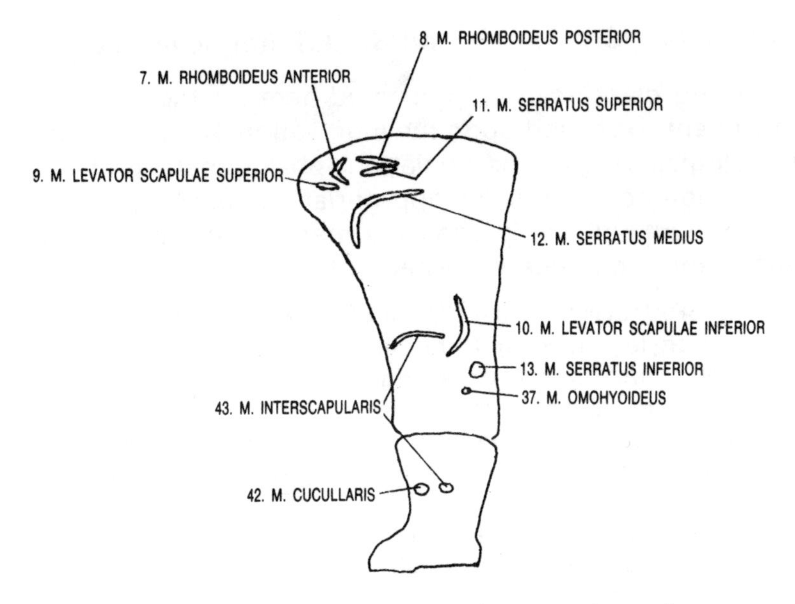

Fig. 25 Internal (medial) view of the right scapula and suprascapula, showing areas of muscle attachment.

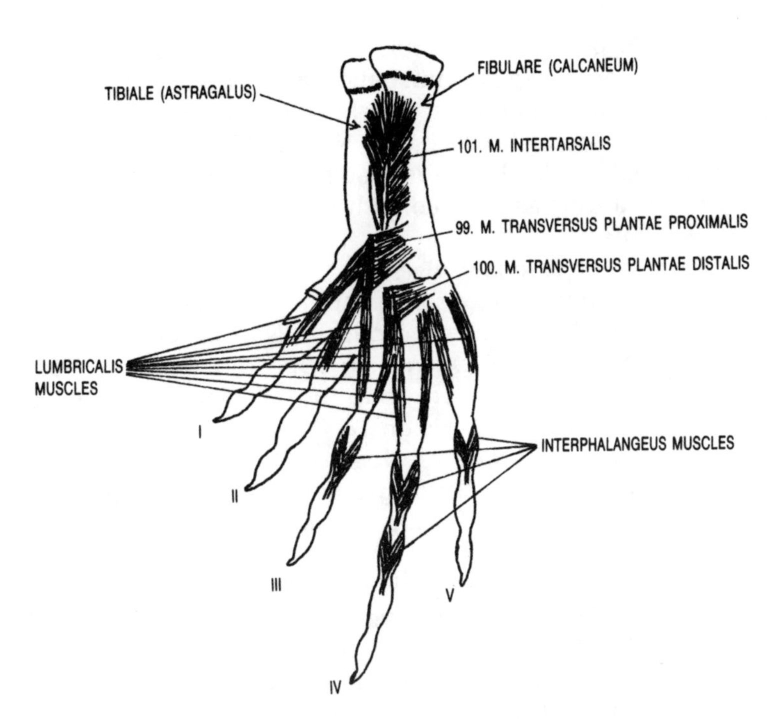

Fig. 26 Deeper muscles on the plantar surface of the tarsus and foot.

J. *A CLASSIFICATION OF THE MUSCLES* (for advanced students only)

The following classification, presented here for the benefit of the more advanced students, is based upon the innervation of the various muscles, on their embryological origin, and on various phylogenetic principles and facts whose discussion does not seem appropriate here. Many of the differences between important muscle groups are stated with an intentional brevity, leaving their more complete explanation to the instructor.

- VISCERAL (BRANCHIOMERIC) MUSCLES—derived embryologically from neural crest, and phylogenetically from gill muscles.
  - Mandibular arch musculature (trigeminal nerve field) Nos. 25–28, 30, 32
  - Hyoid arch musculature (facial nerve field) Nos. 29, 31
  - First branchial arch musculature (glossopharyngeal nerve field) No. 38
  - Posterior branchial arch musculature (vagus and accessory nerve fields) Nos. 39–43
- SOMATIC MUSCLES—derived from somites
  - AXIAL MUSCLES—derived directly from somites; primitively associated with axial skeleton.
    - Derivatives of the prebranchial somites (extrinsic ocular musculature) Nos. 19–24
    - Derivatives of the postbranchial somites
      - Epaxial musculature (field of dorsal rami) Nos. 1–5
      - Hypaxial musculature (field of ventral rami)
        - Craniocervical division
          - Hypoglossal musculature Nos. 33–37
          - Dorsoscapular musculature Nos. 7–13
        - Thoraco-abdominal division Nos. 14–17
        - Sacrocaudal division Nos. 6, 18
  - APPENDICULAR MUSCLES—derived from limb bud mesenchyme of somatic origin; primitively associated with limbs.
    - FORELIMB
      - Dorsal or extensor division
        - Shoulder extensor group Nos. 44–46
        - Brachial extensor group Nos. 52–53
        - Antebrachial extensor group (n. radialis) Nos. 60–66
        - Manual extensor group (n. radialis) Nos. 67–68
      - Ventral or flexor division
        - Pectoral group Nos. 47–51
        - Antebrachial flexor group (n. ulnaris) Nos. 54–59
        - Manual flexor group (n. ulnaris) No. 69
    - HIND LIMB
      - Proximal musculature (hip and thigh)
        - Field of n. ischiadicus Nos. 70, 72–73, 75–80, 83–84, 86–89
        - Field of n. cruralis Nos. 71, 74, 81–82, 85
      - Distal musculature (lower leg and foot)
        - Field of n. tibialis—flexor muscles Nos. 90–91, 96–102
        - Field of n. peroneus—extensor muscles Nos. 92–95, 103–104

CHAPTER 4

Major Internal Features

A. *ORAL CAVITY*

Pry open the oral cavity slightly and insert a pair of scissors. Cut through the jaw articulation (joint) on both sides, thereby extending the mouth in both directions; *do not cut any further posterior than about the level of the center of the tympanum*. (If you have already studied the muscles, you should recognize that you are cutting through the masseter, temporal, and pterygoid muscles, and probably part of the depressor mandibulae as well; avoid cutting any further, however.) Now, force the mouth open with your hands, and, while holding it open, locate and study the following:

1. The **maxillary teeth**, on the margins of the upper jaw. These are very small and numerous, and may easily be felt by inserting your finger into the mouth and pressing against the roof of the mouth as you withdraw it. Examine the structure of these teeth, which are borne by the maxilla and premaxilla (two bones of the upper jaw).
2. The **internal nares**, anterolaterally located on the roof of the mouth. These represent the opening of the nostrils into the oral cavity. Insert a probe through the internal nares and verify their continuity with the external nares. The passage between internal and external nares is called the **nasal passage** or **choana** (plural, *choanae*).
3. The **vomerine teeth**, just medial to the internal nares. A short series of these teeth are borne by each of the vomers, which are bones of the skull (Chapter 2). Examine the structure of these teeth, and compare them to the maxillary teeth.
4. The **eustacian tubes**, which open in the posterolateral corner of the roof of the mouth. These tubes lead to the middle ear; verify this fact by inserting a probe in the eustacian tube and feeling it beneath the tympanum.
5. The **tongue**, anteriorly situated on the floor of the mouth. Study this organ, particularly its mode of attachment. While holding the upper

jaw with one hand, depress the lower jaw with the other, and observe the way in which the posterior margin of the tongue flips up. The frog, when alive, catches insects by performing a similar motion at high speed.

6. Two very small, porelike openings to the **vocal sacs**, in the posterolateral corner of the floor of the mouth. The vocal sacs are difficult to find and are not always present; they are more often lacking in females than in males.
7. The **glottis**, an opening located far posteriorly on the ventral midline of the oral cavity. The glottis is more often slitlike in the male than in the female; this difference reflects the fact that only the males emit the characteristic mating calls to attract the females. These mating calls or "croaks" differ from one species to the next, each species having its own peculiar sound.
8. The **larynx**, or voice box, lies just beneath the glottis. The *laryngeal cartilages* which support it are described elsewhere (Chapter 2).
9. The entrance to the **esophagus**, at the posterior end of the oral cavity.
10. The paired **thyroid glands** (advanced students only), located to either side of the hyoid cartilage just posterior to the posterior horns. This gland secretes in the frog an iodine-rich hormone which controls its metamorphosis from a tadpole.

B. GENERAL BODY CAVITY

If you have not yet studied the muscles, you must first remove the skin from your frog's ventral surface. Pierce the skin (but not the underlying abdominal muscles) and make a longitudinal incision along the midventral line, from a point between the forelimbs to a point between the hind limbs. Make two transverse incisions, one at either end of your first incision. Fold back laterally the two flaps of skin thus formed, exposing the abdominal muscles.

The shiny, white line now visible along the ventral midline is called the **linea alba** (Latin: "white line"). The longitudinally running muscle fibers, parallel to the linea alba, and within the first centimeter or so adjacent to it on either side, belong to the *rectus abdominis* muscle.

To expose the body cavity, pierce the rectus abdominis muscle just to one side of the ventral midline. With a pair of scissors, lift up the abdominal muscles as you cut through them carefully, just to one side of the linea alba. Be careful that the point of the scissors does not go deep and injure the underlying structures; be especially careful not to injure the heart. Extend this longitudinal incision anteriorly until the heart is exposed (you may have to cut through part of the shoulder girdle), and posteriorly to the level of the hind limbs. Locate a large vein, the **ventral abdominal vein**, adhering to the deep side of the linea alba along its length; gently free this vein from the abdominal layer of muscles, being careful not to damage it. Then, and only then, should you attempt to make two transverse incisions, one just anterior to the heart, and the other as far posteriorly as you can, being careful not to injure the ventral abdominal vein or any other structure. Extend both these transverse incisions around to the dorsal side of the animal; peel back the flaps and expose the general body cavity.

The general body cavity or **coelom** is lined throughout with a membrane known as the **peritoneum**, and is therefore sometimes called the **peritoneal cavity**. The peritoneum also lines all organs located in the body cavity; such organs are generally called **viscera**. When the peritoneum lining an organ is continuous with the peritoneum lining the body cavity as a whole, a double-layered **mesentery** is formed. Such mesenteries may arise in either of two ways, as shown in Fig. 27: Organs derived from the gut itself (stomach, intestine, etc.) become enveloped from both sides as the right and left coelomic cavities grow together; they are originally connected by both a dorsal and a ventral mesentery, although the latter usually disappears. Organs originally *retroperitoneal* in position (i.e. behind the peritoneum) may bulge into the body cavity, carrying part of the lining of this cavity with them. In both cases, note that the mesentery is always a double-layered membrane.

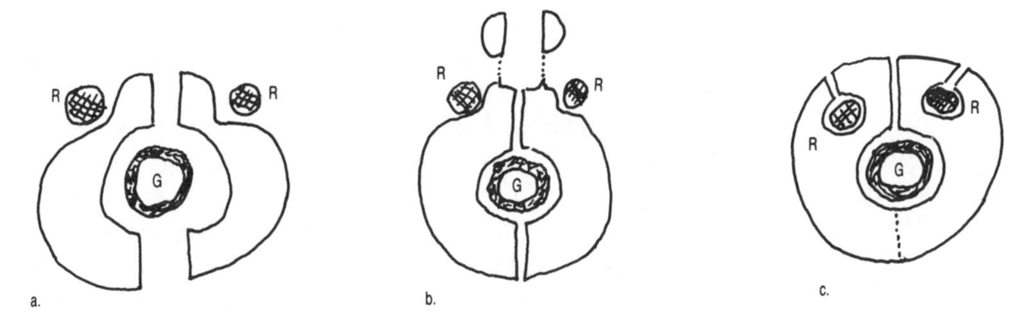

Fig. 27 The development of mesenteries in an organ (such as the stomach) which is part of the gut (G), and in a retroperitoneal organ (R) such as one of the gonads. Schematic only; many details omitted.

If your frog is a female collected during the breeding season, her ovaries will be greatly expanded and will fill a considerable portion of the abdominal cavity; numerous black and white **ova** (eggs) will be visible. If the ovaries are so much enlarged that they would greatly interfere with the study of other organs (i.e. if other organs are not visible because of the condition of the ovaries), ask your laboratory instructor for permission to remove them. Be careful not to damage blood vessels or other structures if you should remove the ovaries.

C. VISCERA

The organs of the general body cavity are collectively known as the **viscera**, though this name is sometimes restricted to the digestive organs only. Move the organs around in the general body cavity until you have located and studied the following:

1. The **heart**, which is located in a separate **pericardial cavity**, surrounded by a **pericardial sac** (or **pericardium**) which may have been damaged by the supply house when the arteries were injected. The pericardium separates the pericardial cavity from the remainder of the general body cavity, which is called the **abdominal cavity**.
2. The **liver**, deep to the heart and also more posterior, but superficial (ventral) to the other major structures. The liver is subdivided into three large **lobes** (right, left, and median). It is usually brown in color, but it may become partially filled with blue or yellow later when the veins are injected. The secretion of bile and the storage of glycogen are but two of the liver's many functions.
3. The **gall bladder**, a greenish sac, between the lobes of the liver. The gall bladder stores the bile secreted by the liver; one of the bile pigments is responsible for the green color.
4. The **lungs**, deep to the lobes of the liver, and at the anterior limit of the abdominal cavity. The surface of the lungs appears wrinkled or spongy. During the injection of red latex into the arteries, the lung usually acquires a pink or reddish color (its normal color is light brown). The lungs are connected to the larynx by a pair of **bronchial tubes** or **bronchi**. Probe through the glottis into one of the bronchi to verify this connection.
5. The **stomach**, an enlarged portion of the digestive tract, located just under the left lobe of the liver. ("Left" means the *frog's* left, not *your* left.) Open the stomach with a longitudinal incision; study any half-digested remains you may find. Wash out the stomach, and study its lining. Insert a probe into the stomach, and probe anteriorly through the esophagus; you should be able to see your probe emerge in the oral cavity.

   You may notice that the stomach and certain other digestive organs are connected to the dorsal midline of the body cavity by means of mesenteries. Blood vessels and nerves often run within these mesenteries, but sometimes they may also run free (unattached). Be careful not to damage either the mesenteries or the blood vessels and nerves as you dissect.
6. The **small intestine**, which consists of a duodenum and an ileum. The **duodenum** is the wider, less convoluted portion immediately following the stomach; the lower portion of the stomach, adjacent to the duodenum, is called the **pylorus**, and the constricted portion between

the two organs is known as the **pyloric valve**. The **ileum** is the longer, narrower, and more convoluted portion of the small intestine.

7. The **large intestine**, a relatively short tube, extending from the small intestine to the cloaca. The large intestine is of greater diameter than even the duodenum.
8. The **cloaca**, at the posterior end of the digestive tract. Since the cloaca also receives the excretory and reproductive ducts, it is the common terminal portion of all three systems. The opening of the cloaca to the outside is called the **anus**.
9. The **pancreas**, a long, thin mass of light-colored, glandular tissue lying alongside and parallel to the duodenum and the lower (pyloric) portion of the stomach. It is attached to the same mesentery as these parts of the digestive tract, and it may be seen easily by lifting the lower end of the stomach and looking underneath. In addition to secreting enzymes which aid in digestion, the pancreas also secretes the hormone **insulin**.
10. The **bile duct**, which brings material from the liver, gall bladder, and pancreas into the duodenum. The individual portions of this duct have been given separate names: the **pancreatic duct** drains the pancreas, the **hepatic duct** drains the liver, and the **cystic duct** (probably the easiest to find) drains the gall bladder. The portion common to all three is often called the *common bile duct*, which empties into the duodenum. Search carefully for this system of ducts and examine the various tributaries.
11. The **spleen**, a spherical organ located in the mesentery which lies between the duodenum and large intestine, on the left side of the frog. It usually fills with blue latex during the injection of the veins; in well-injected specimens it may be entirely blue. (Its original color is a very dark reddish-brown.) The spleen removes and decomposes old blood cells; some of the breakdown products are eventually secreted as bile pigments by the liver.

Although you will dissect these at a later time, you may already have noticed certain other structures, belonging to the urogenital system: the *gonads* (ovaries or testes), *oviducts* (folded tubes in females only), *fat bodies* (yellowish, with fingerlike projections), *kidneys*, and *urinary bladder*. You need not search for these organs now; their descriptions are included in Chapter 6. Be careful, however, not to mistake the oviducts for a piece of intestine!

D. MICROSCOPIC SLIDES

If the necessary material is available, examine prepared slides of the various internal organs listed above. Your laboratory instructor should provide you with a list of the slides available for study.

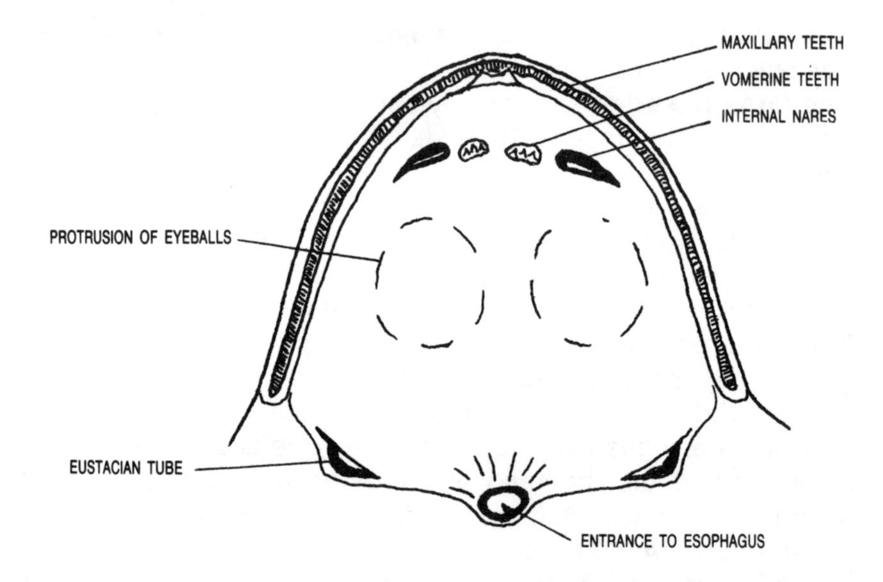

Fig. 28 Roof of the mouth.

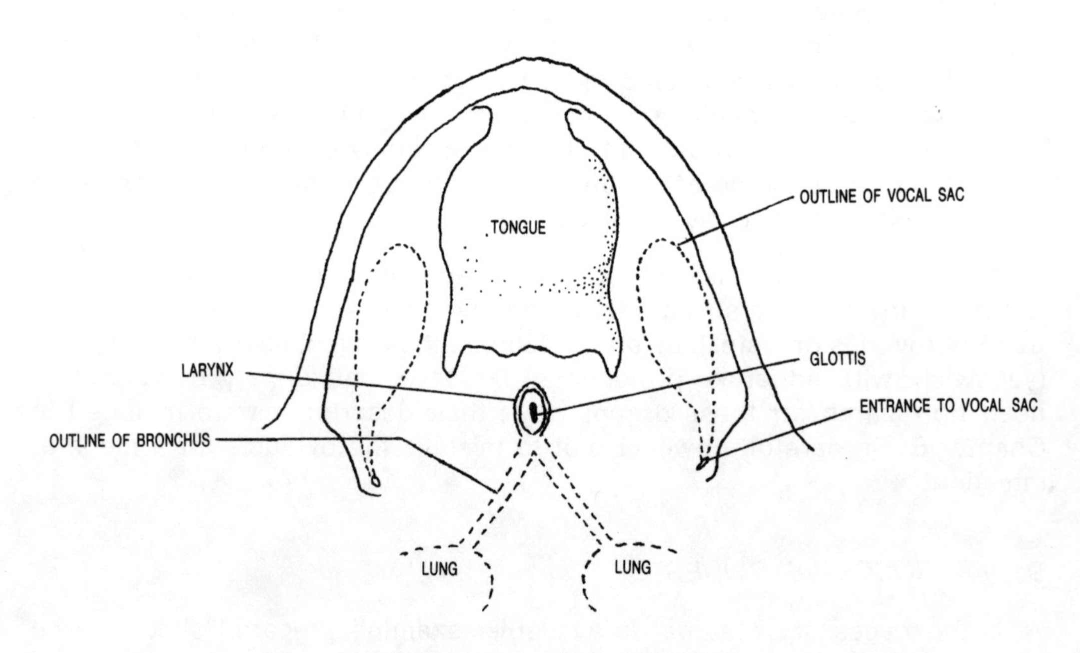

Fig. 29 Floor of the mouth.

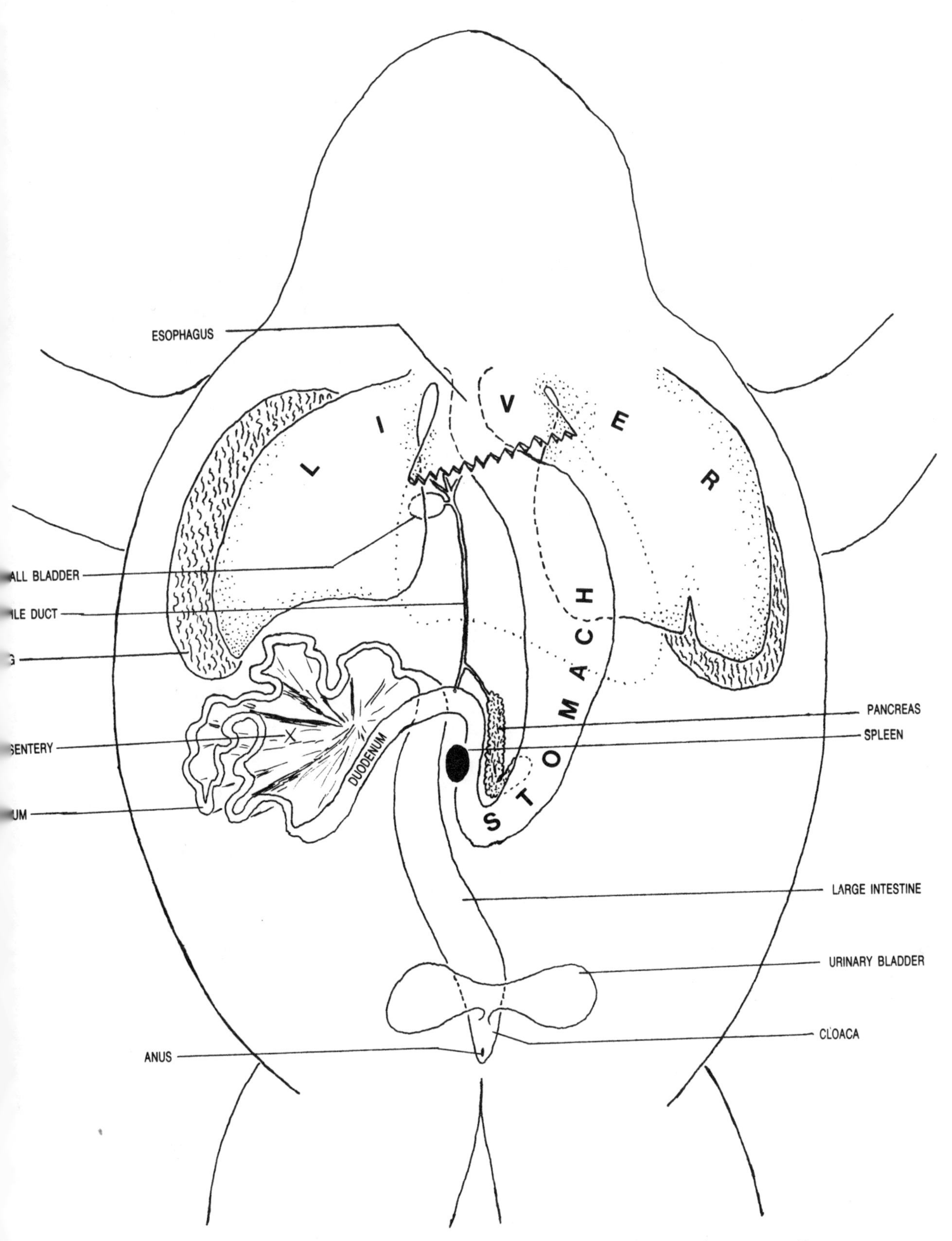

Fig. 30 Major internal organs (middle lobe of liver mostly removed).

CHAPTER 5

Circulatory System

In order to make the blood vessels more easily visible, the arteries and veins are usually injected with colored latex. Doubly injected frogs usually have red latex in their arteries and blue latex in their veins; triply injected frogs usually have the arteries in red, the pulmonary and systemic veins in blue, and the renal and hepatic portal systems in yellow. In order to make these injections, your frog was probably cut by the supply house in at least two places: across the chest region, in order to reach the heart, and along the thigh, in order to reach the sciatic vein.

(Advanced students may wish to note that the arteries are injected in the direction of blood flow, beginning at the heart; this injection is usually performed first, to prevent the filling of the arteries with blue latex during the venous injection. The pulmonary and systemic veins are then injected in the direction opposite to blood flow, beginning at the sinus venosus. This injection must be performed with sufficient force to break through the valves; often the walls of the veins rupture instead, and the body cavity begins to fill with the latex. The portal systems are injected through the sciatic vein, largely in the direction of blood flow. Collateral circulation in the leg fills the femoral vein. The hepatic portal vein and its tributaries must fill in a direction opposite to that of blood flow; the filling of these vessels is often poor as a result.)

All students should be careful to notice that, while arteries *branch* into finer and finer blood vessels, veins can more often be described as comparable to a system of rivers, with smaller veins (like smaller streams) being *tributaries*, not branches, of the larger ones. (The portal veins are exceptional, since they do branch.) Also notice that arteries *supply* blood *to* the organs which they serve, while veins *drain* blood *from* these organs.

The blood vessels, particularly the veins, are rather variable. The pattern of branchings and regroupings described below may differ from that observed in your own frog. (In particular, it is not uncommon at all for the tributaries to a vein to joint in a different order; it is a bit less common for the branches of arteries to arise in a different order.)

The arteries, particularly the larger ones, are circular in cross section and their walls are thick. The veins, by contrast, may often have irregular cross sections which conform to fill the space between surrounding organs. The walls of the veins are thin and latex is always visible through them. Uninjected veins tend to collapse, while uninjected arteries usually retain their shape. Since the flow of venous blood is usually slower, a vein draining an organ will generally have a larger cross section than the artery supplying that same organ. If the necessary slide material is available, examine and compare microscopic cross sections of arteries and veins. You may also at this time wish to examine slides of various types of blood, blood-forming (hemopoietic) tissues, and lymph.

A. HEART

If you have not already done so, cut through the ventral body wall, including the shoulder girdle, and expose the heart. The heart is surrounded by a thin, membranous **pericardium** (Greek: *peri*-, "around" or "surrounding"; *cardia*, "heart"), which you have probably already cut.

Move the heart from side to side and observe the **sinus venosus** behind it on the dorsal wall of the pericardial cavity. The sinus venosus is formed by the two superior venae cavae, which approach from the sides, and the inferior vena cava, which approaches from further posteriorly. These three veins give a triangular shape to the sinus venosus. Blood flows from the sinus venosus into the right atrium of the heart. The opening is guarded by a paired valve, the **sinoatrial valve**, which prevents backflow.

The heart itself consists of two **atria** (singular, *atrium*; also formerly called *auricles*) and a single **ventricle**. The right atrium receives blood from the sinus venosus. The left atrium receives blood from the lungs. From the two atria, blood enters the single ventricle through the **atrioventricular valve**. During contraction of the ventricle, this valve closes to prevent backflow.

The **ventricle** is the conical, posteriorly directed chamber of the heart. Since it must propel the blood throughout the rest of the body, the ventricle possesses a thick, muscular wall. The **semilunar valves** guard the passage from the ventricle to the truncus arteriosus. The **truncus arteriosus** (or *conus arteriosus*) is a stout, cylindrical vessel which carries the blood away from the heart.

B. ARTERIES

Upon leaving the heart, the truncus arteriosus runs anteriorly, branching into three pairs of arteries: the right and left **pulmocutaneous arteries** (the more posterior pair), the right and left **systemic arteries** (the middle pair), and the right and left **common carotid arteries** most anteriorly. Each common carotid immediately splits into separate **internal** and **external carotid arteries**. These first few branchings occur in rapid succession, and may frequently vary: the systemic and common carotid arteries may be conjoined, for example, and the distance covered by the common carotid before its first branching may be long, short, or nonexistent.

1. The **pulmocutaneous artery**, arising from the truncus arteriosus, soon branches into a more anterior **cutaneous artery** and a more posterior **pulmonary artery** on either side of the body. The pulmonary arteries supply blood to the lungs; the cutaneous arteries ramify just beneath the surface of the moist skin. The exchange of gasses (oxygen, carbon dioxide) takes place both through the skin and through the alveoli of the lungs. (Advanced students should follow one of the cutaneous arteries, and observe that it branches into **dorsal, lateral**, and **anterior rami**. The anterior ramus gives off a **ramus auricularis** to the tympanic region, and finally communicates with the temporal artery, described below, to send blood into the superficial parts of the face.)
2. The **external carotid artery** is the more medial and more superficial of the carotid arteries. It supplies blood to the tongue and the floor of the mouth. (Advanced students should follow the external carotid anteriorly, where it gives off many small branches to the muscles surrounding the hyoid apparatus, finally continuing as the **lingual artery** beneath the tongue.)
3. The **internal carotid artery** is somewhat larger than the external carotid and is situated more laterally and also deeper. The internal carotid, or sometimes the common carotid, has a bulblike swelling at its base, the so-called **carotid bulb**. This bulb swells and contracts elastically to insure that the brain and other head structures receive a more steady, even blood pressure, rather than a pulsating one. (Advanced students should follow the internal carotid as it runs anteriorly from the carotid bulb to enter the head, giving off first a **cerebral artery** to the brain, and an **ophthalmic artery** to the eye region; the ophthalmic artery may arise either from the cerebral artery, or directly from the internal carotid. The cerebral artery also gives off a **ramus posterior**, which runs posteromedially to meet with its partner of the opposite side to form the median, unpaired **basilar artery**, which supplies blood to the spinal column. Beyond the cerebral and ophthalmic arteries, the internal carotid continues as the **palatine** artery, supplying blood to the roof of the mouth, and giving off a branch which forms a loop reaching to the maxillary artery; a **palatonasal artery** arises from this loop.)
4. The **systemic arteries** arise from the truncus arteriosus. After splitting away from the carotid arteries, the systemics turn laterally and then dorsally to form the right and left **aortic arches**. Each aortic arch gives

off, in succession, a small **laryngeal artery**, a large **occipitovertebral artery**, and a larger **subclavian artery**. (Advanced students should locate the very small **laryngeal artery**, which arises just beyond the carotid arteries. It supplies blood to the larynx and the anteriormost part of the esophagus.)

5. The **occipitovertebral artery** (for advanced students only) supplies blood to the esophagus, the head and jaws, the dorsal body wall, the abdomen, and the brain (via the vertebral column). The **esophageal artery**, to the esophagus, is given off first. Just beyond this branch, the remainder of the occipitovertebral artery bifurcates into a more anterior **occipital artery** and a more posterior **dorsovertebral** artery.

   The occipital artery runs into the head, where it branches into an **orbitonasal artery** and a somewhat larger **temporal** artery. The latter arches ventrolaterally to form a loop which finally unites with the ramus anterior of the cutaneous artery. From this arch or loop arise the **internal mandibular artery** to the inner margins of the lower jaw, the **external mandibular artery** to the outer margins of the same, and the **maxillary**, or **superior maxillary artery**, along the margins of the upper jaw.

   The **dorsovertebral artery** gives off a **ramus cranialis**, or **vertebral artery**, which runs dorsally toward the vertebral column to meet the basilar artery along the dorsal midline. This provides another route by which blood may reach the brain. The remainder of the dorsovertebral artery continues along the dorsal side of the body cavity, paralleling the vertebral column, and supplying a series of **abdominal arteries** to the muscles of the abdominal wall. Smaller branches are also given off medially to the muscles of the spinal column and adjacent dorsal regions; some of these branches communicate with the lumbar arteries in the lumbar region.

6. The **subclavian artery** supplies the muscles of the shoulder and forelimb; trace it to a point just beyond the shoulder joint, where it enters the arm. (Advanced students should continue further, following the subclavian artery into the arm, where it becomes the **brachial artery**. The brachial artery bifurcates just proximal to the elbow into separate **radial** and **ulnar arteries**, running parallel to one another along the forearm.)

The **dorsal aorta** is formed by the union of the right and left aortic arches. As it courses posteriorly, it gives off the following branches (Nos. 7–10), in order:

7. The **coeliacomesenteric artery** (unpaired), which divides into a coeliac and an anterior mesenteric artery. The **coeliac artery** divides further into a **hepatic artery** to the liver, and a **gastric artery** supplying the stomach and pancreas. The **anterior mesenteric artery** gives off a

splenic or **lienic artery** to the spleen (Latin name, *lien*), and then bifurcates into an **intestinal artery**, to the small intestine, and a **hemorrhoidal artery**, to the large intestine.

8. The **urogenital arteries** (six pairs), which supply the kidneys, gonads, and fat bodies. All six pairs may be termed **renal arteries** (*ren* means kidney in Latin), but only the first pair or two give off **genital** (ovarian or spermatic) **arteries** to the gonads.
9. The **lumbar arteries** (several pairs), which originate along much of the length of the dorsal aorta, especially in the lumbar region (the "small of the back"). These arteries supply the dorsal body wall, the segmental muscles, and the vertebral column; they also communicate with branches of the dorsovertebral artery.
10. The **posterior mesenteric artery** (unpaired), which arises from the dorsal aorta at the posterior end of the coelom. It supplies the posterior end of the large intestine, and also the cloaca.
11. The dorsal aorta now bifurcates, forming a pair of **common iliac arteries**. Each common iliac gives rise to two branches (hypogastric and epigastric arteries) before itself bifurcating. The **hypogastric artery** (paired) supplies the urinary bladder; the **epigastric artery** (paired) supplies the ventral body wall.

    Each common iliac artery then bifurcates, continuing as a smaller femoral and a larger sciatic artery. The **femoral artery** (**external iliac artery**) runs along the outer side of the thigh, and supplies the extensor muscles (basically) of this region. The **sciatic artery** (**internal iliac artery**) runs along the posteromedial margin of the thigh, supplying not only the flexor musculature (generally) of this region, but also the entire shank and foot.

C. VEINS

1. Pulmonary Circulation

The **pulmonary veins** return freshly oxygenated blood from the lungs to the left atrium of the heart. They usually join together first and then enter the heart as a single vessel.

2. Systemic Circulation

This is more readily studied by tracing the veins backward from the sinus venosus.

The **anterior vena cava**, or **precaval vein** (paired) drains blood from the anterior regions of the body, including the head and forelimbs. Its three tributaries, all large, are:

(a) The **external jugular vein**, the most anterior of the three, formed by the confluence of the lingual and mandibular veins. The **lingual vein** drains the tongue and the floor of the mouth; the **mandibular vein**, located more laterally, drains the blood from the region of the lower jaw.
(b) The **innominate vein**, formed by the confluence of the subscapular and internal jugular veins. The **internal jugular vein** drains blood from the brain; the **subscapular vein** drains blood from the shoulder and part of the forelimb.
(c) The **subclavian vein**, the most posterior of the three, formed by the confluence of the brachial and musculocutaneous veins. The **brachial vein** drains blood from the arm. The **musculocutaneous vein** drains blood from the muscles on the lateral side of the trunk in the shoulder region, and also from the many subcutaneous vessels in the region supplied by the cutaneous artery. From the head and shoulder regions, the musculocutaneous artery runs posteriorly to a point about midway along the body cavity, then it reverses its direction to run anteriorly and join the subclavian.

The **posterior vena cava**, or **postcaval vein** (unpaired) runs anteriorly along the dorsal wall of the body cavity. It is formed by the confluence of the four pairs of **renal veins**, which drain blood from the kidneys. It also receives a single pair of **genital** (ovarian or spermatic) **veins**, which drain blood from the fat bodies as well as from the gonads. Shortly before entering the sinus venosus, the posterior vena cava also receives blood from the prominent **hepatic veins** (paired), which drain the liver (Latin name, *hepar*).

3. Renal Portal System

The renal portal system is best studied beginning with the sciatic vein at the place it was injected. The **sciatic vein** (paired, as are all vessels in the renal portal system) drains blood from the posterior or flexor surface of the leg. At the base of the thigh, a **vena communicans** is formed between the sciatic and femoral veins. More anteriorly, the sciatic and femoral veins join to form the **renal portal veins** of either side, which enter the substance of the kidneys. Just anterior to the union of the sciatic and femoral veins, a **dorsolumbar vein** also enters the renal portal vein.

The **femoral vein** returns blood from the anterior or extensor side of the leg. It twists around the anterior side of the leg, and enters the body cavity near the ischium. It gives off, just anterior to the vena communicans, a **pelvic vein**, which belongs to the hepatic portal system (see below). The femoral vein then runs anteriorly to join the sciatic vein.

4. Hepatic Portal System

The hepatic portal system consists of two largely independent sets of vessels: the ventral abdominal vein (and its tributaries), and the hepatic portal vein (with its tributaries).

The **pelvic veins** (paired) arise from the femoral veins just anterior to the vena communicans. Each pelvic vein receives as a tributary a small **vesicular vein**, draining the urinary bladder. The two pelvic veins meet each other and unite to form the **ventral abdominal vein** (unpaired). This prominent vein was observed during the initial inspection of the viscera; it runs along the ventral midline of the body cavity. Anteriorly, the ventral abdominal vein splits into right and left halves, each entering its respective lobe of the liver. The branch to the left lobe may join with the hepatic portal vein before entering the liver. A small **pericardial vein** (unpaired, flowing posteriorly) may join the ventral abdominal at or near the point of branching; this vessel drains the pericardial cavity and the heart muscle.

The remainder of the hepatic portal system consists of the **hepatic portal vein** and its tributaries, all unpaired, draining blood from the major digestive organs. The major tributary is the **intestinal vein**, formed by the union of the **splenic** or **lienic vein**, draining the spleen, the **hemorrhoidal veins**, draining the large intestine and cloaca, and the **mesenteric veins** from the ileum of the small intestine. The intestinal vein runs anteriorly, receiving two additional tributaries: a **duodenal vein** from the pancreas and duodenum, and a **cystic vein** from the gall bladder. The intestinal vein then merges with the **gastric vein**, from the stomach, to form the hepatic portal vein. The latter enters the substance of the liver, usually together with the left half of the ventral abdominal vein.

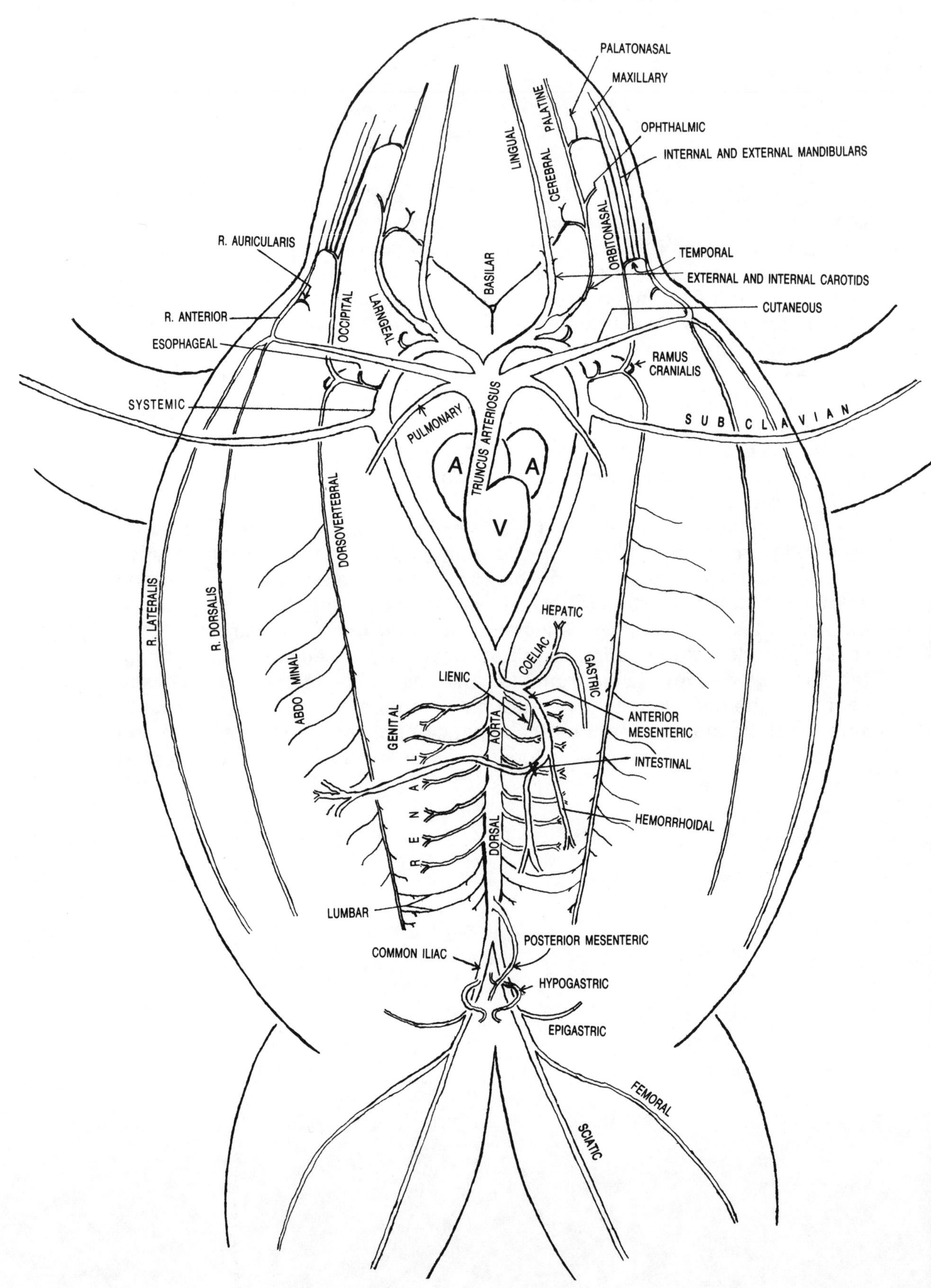

Fig. 31 Arteries of the frog, ventral view.

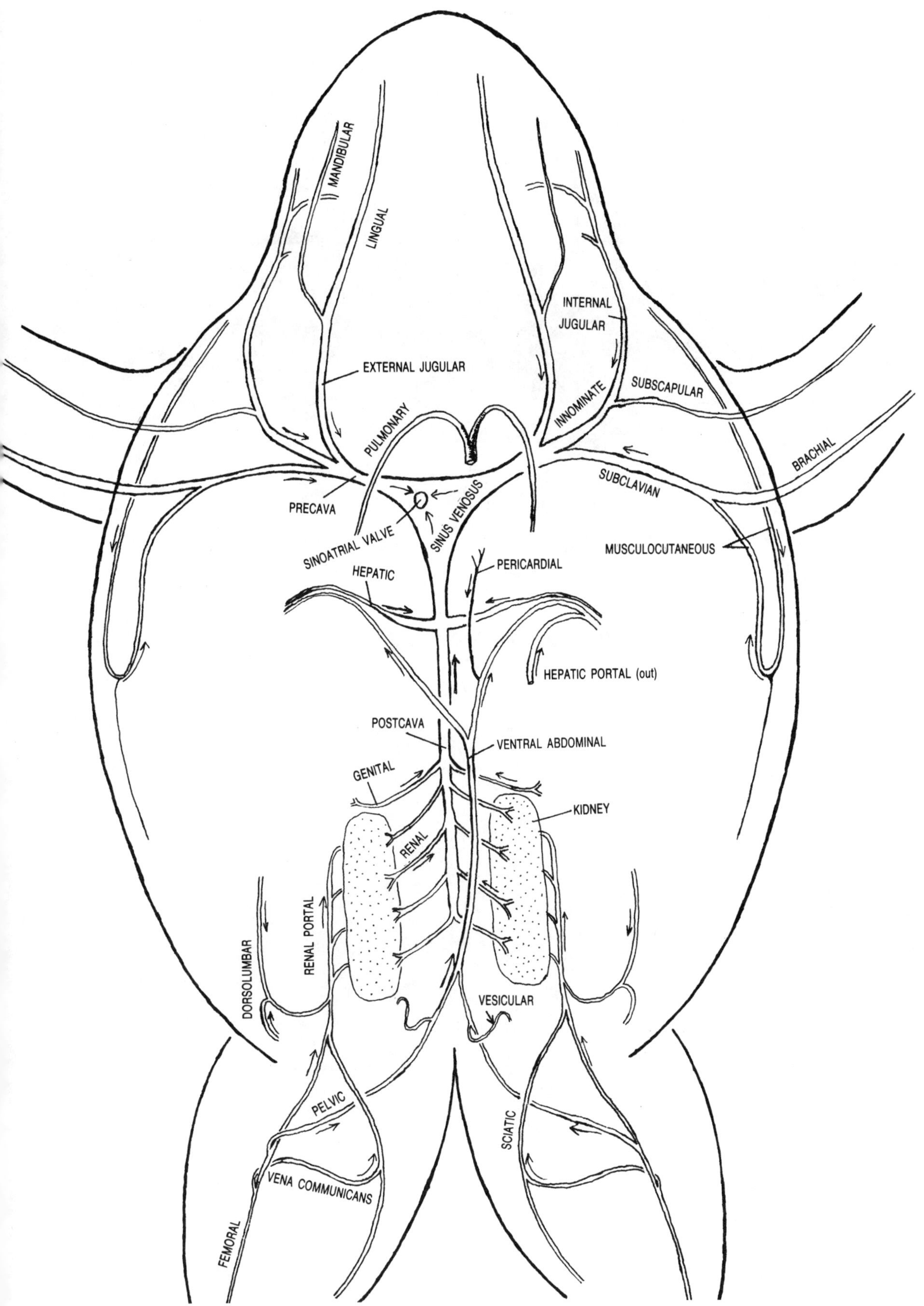

Fig. 32 Veins of the frog, ventral view, with heart and viscera removed. The tributaries of the hepatic portal vein are shown separately in Fig. 34.

OPHTHALMIC
ORBITONASAL
R. AURICULARIS
OCCIPITAL
R. DORSALIS
R. LATERALIS
CEREBRAL
39-41. MM. PETROHYOIDEI POSTERIORES I, II, AND III
INTERNAL CAROTID
TEMPORAL
PALATINE
SYSTEMATIC
R. ANTERIOR
PALATONASAL
CUTANEOUS
38. M. PETROHYOIDEUS ANTERIOR
MAXILLARY
EXTERNAL MANDIBULAR
INTERNAL MANDIBULAR
PULMONARY
PULMOCUTANEOUS
EXTERNAL CAROTID
LINGUAL
VENTRICLE
COMMON CAROTID
CAROTID BULB
ATRIA
TRUNCUS ARTERIOSUS

Fig. 33 Arterial circulation to the head region, lateral view.

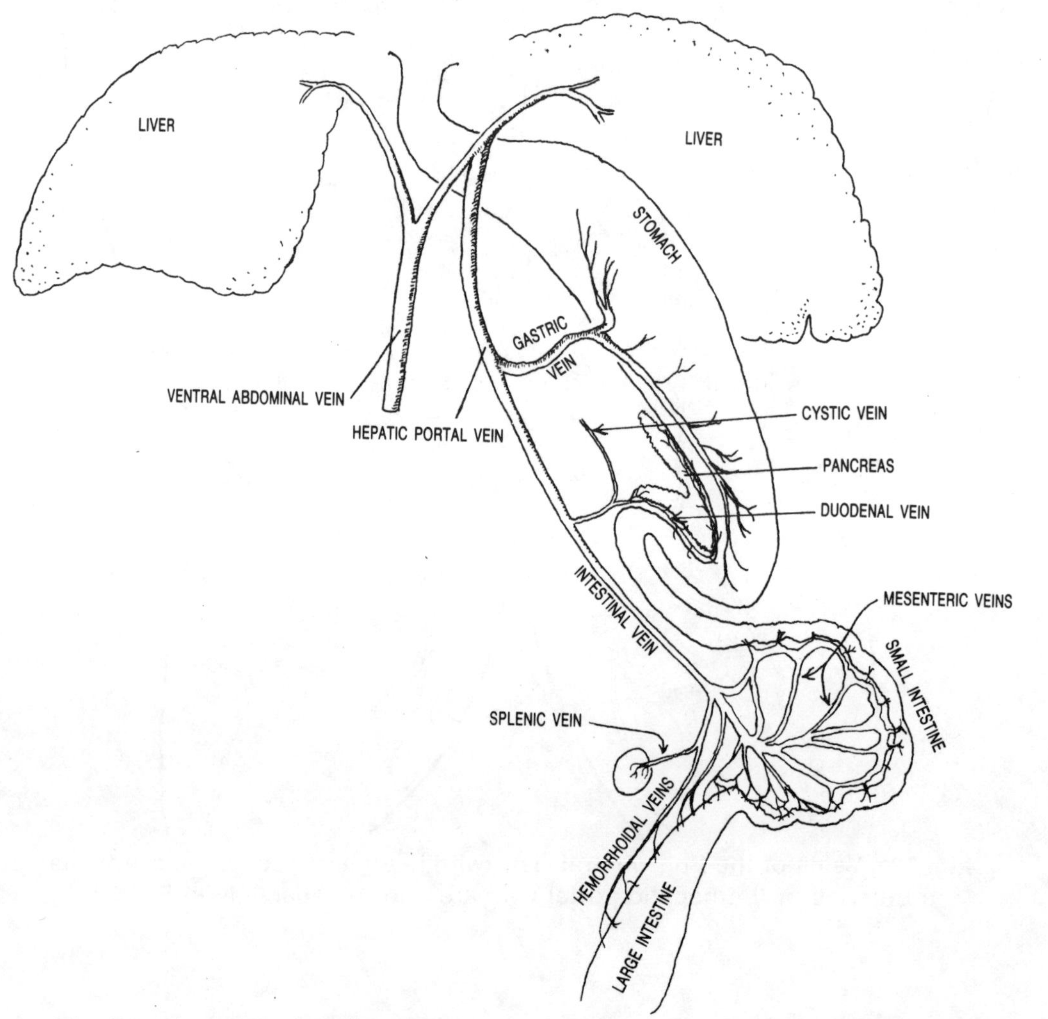

Fig. 34 Hepatic portal circulation. Ventral view, with viscera pulled aside.

CHAPTER 6

Urogenital System

A. URINARY OR EXCRETORY SYSTEM

The major organs of the excretory system are the paired **kidneys**. The kidneys are located dorsal to all other organs lying in the abdominal cavity. They are even dorsal to the peritoneum itself, and are thus **retroperitoneal** in position. The kidneys are elongated structures, originally reddish-brown, but usually blue in specimens where the veins have been injected with latex. Although they are not part of the urogenital system, the **adrenal glands** may be noted at this time as a slender, elongated mass of yellowish tissue adherent to the ventral surface of the kidneys.

The urine is collected from the kidneys by the **archinephric ducts** (**Wolffian ducts**), which are often incorrectly called ureters. These ducts originate on the lateral margin of each kidney, from which they course posteriorly and finally empty into the cloaca. The archinephric duct in the female carries urine only; in the male, the archinephric duct carries both the urine and the sperm.

The **urinary bladder** is a thin-walled, expandable sac, lying ventral to the cloaca. It is generally dumbbell-shaped, with expanded lateral portions; its connection to the cloaca arises from the constricted middle portion.

B. REPRODUCTIVE SYSTEM

Study first the reproductive system in your own frog. Then switch specimens with someone whose frog is of the opposite sex. You are responsible for the reproductive systems of both sexes.

Male Reproductive System

The **testes**, or male gonads, are a pair of yellowish, ovoid bodies resting upon the ventral surface of the anterior portion of the kidneys. The testes are truly retroperitoneal organs, but they have pushed their way into the abdominal cavity, carrying with them a **mesentery**, or fold of peritoneum (see Fig. 27). The mesentery which supports the testes is known by the special name of **mesorchium**. The **fat bodies** (yellowish, with many fingerlike projections) develop at the anterior margin of the mesorchium.

Gently pull the testes medially, and find the **vasa efferentia**, or efferent ductules, which are a series of fine tubules carrying the sperm from the testes to the kidneys. The sperm passes through the kidneys and then into the archinephric duct. Note that in the male, the archinephric ducts are expanded to form **seminal vesicles** just before emptying into the cloaca.

Note that the male frog possesses no copulatory structures. Fertilization takes place externally; the male deposits his sperm over the eggs of the female, which have been laid, unfertilized, in the water.

Female Reproductive System

The **ovaries**, or female gonads, are immediately recognizable by the presence within them of numerous black-and-white, spherical ova. The ovaries vary greatly in size; during the mating season (in the spring) they become greatly enlarged and occupy a considerable portion of the abdominal cavity. Like the testes, the ovaries are retroperitoneal organs which have pushed their way into the abdominal cavity, carrying with them a **mesentery**, or fold of peritoneum (see Fig. 27). The mesentery which supports the ovaries is known as the **mesovarium**; along its anterior margin, the **fat bodies**, described above, may develop.

The **oviducts** are slender, coiled ducts that may at first be confused with the intestine. However, the oviducts are usually more slender and their folds are much stronger than those of the intestine. The oviducts are also located dorsal to the ovaries. Follow one of the oviducts anteriorly, and find the **ostium**, a funnel-shaped opening into which the ova enter. The ova are shed from the ovaries into the abdominal cavity, where they find their way into the ostia and pass through the oviducts. Shortly before it opens into the cloaca, the oviduct possesses a slightly expanded portion which is often incorrectly termed a **uterus** (the frog has no true uterus). Here, the oviduct crosses superficial to the archinephric duct and empties into the cloaca just anteriorly to the archinephric duct.

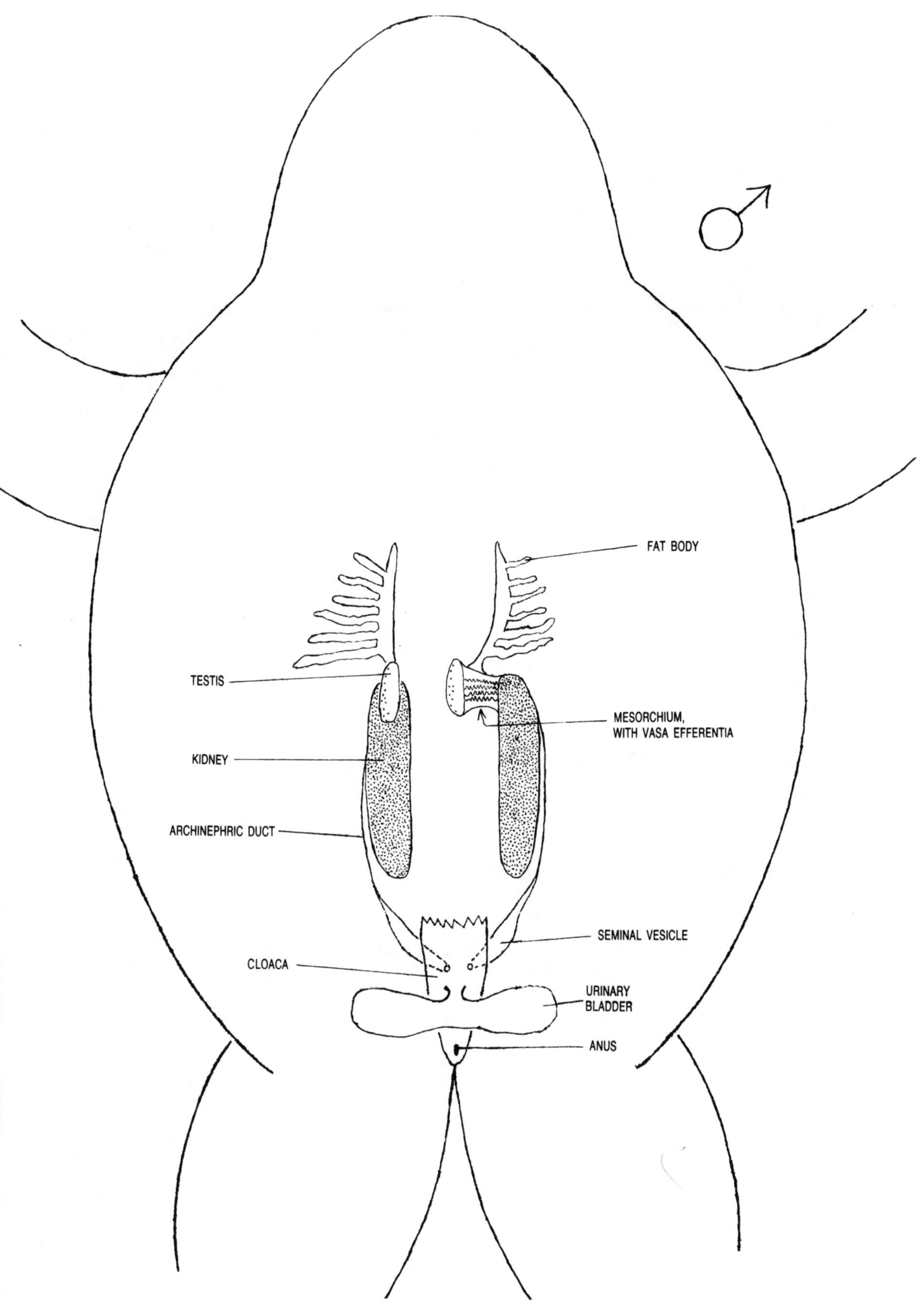

Fig. 35 Male urogenital system, ventral view. The left testis has been reflected medially to reveal the mesorchium and the attachment of the fat body.

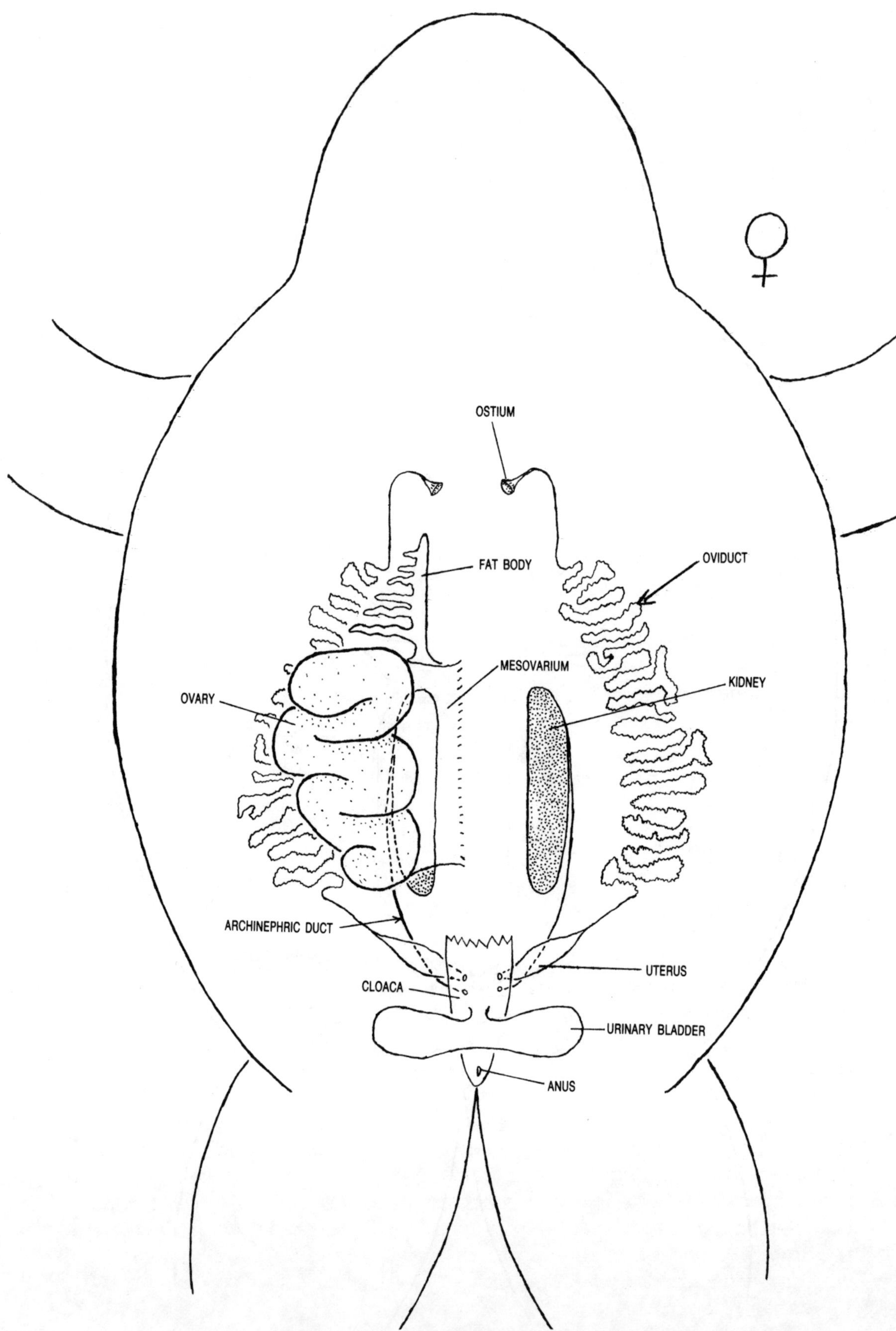

Fig. 36 Female urogenital system, ventral view, with left ovary and fat body removed.

CHAPTER 7

Nervous System

CENTRAL NERVOUS SYSTEM

The dissection of the central nervous system is difficult. The central nervous system, especially the brain, is well protected by a bony surrounding (braincase and vertebral column) which can only be removed by rather brute methods. These methods are likely to damage the nervous tissue itself, which is very delicate. Thus, only those who are skillful in dissection may expect good results.

A. BRAIN

Cut or scrape the temporalis muscles away from their origin on the braincase. If possible, try to obtain a pair of bone-cutting shears. Working forward from the posterior end of the braincase, carefully crush the dorsalmost portion only of the braincase, leaving the lateral walls of the braincase intact. Remove the broken pieces with blunt forceps (be careful to avoid damage to the brain within!), and repeat the procedure until the dorsal surface of the brain is exposed. *Do not use scissors or scalpels*; these instruments will be severely dulled by such use and will not produce good results.

Advanced students should also remove the lateral walls of the braincase in preparation for the removal of the brain itself. Use either blunt forceps or long-nosed pliers to break away the bone piece by piece. Be especially careful in the region of the medulla not to damage any of the cranial nerves. Move the brain from side to side in order to expose these cranial nerves, and delicately cut off each cranial nerve as far from the brain as you can. When you have freed the brain, you should be able to lift it delicately and remove it to a small dish of water or formalin or alcohol for observation under the dissecting microscope.

Identify the following portions of the brain:

1. **Olfactory lobes**, most anteriorly, separated by a shallow median furrow. If you have removed the brain, study also the olfactory tract on the ventral surface of the olfactory lobes. The olfactory lobes and olfactory tracts are concerned with the sense of smell.
2. **Cerebral hemispheres**, just posterior to the olfactory lobes and separated from each other by a more pronounced median furrow. The cerebral hemispheres of the frog are still largely concerned with their primitive function of coordinating different olfactory sensations and discriminating various odors. (Only in mammals do the cerebral hemispheres acquire the "higher" centers of learning that we usually tend to associate with them; the hemispheres of frogs are accordingly many times smaller than in any mammal.)
3. The **diencephalon**, visible from the dorsal side as a diamond-shaped, depressed area between the cerebral hemispheres and the optic lobes. The **pineal body** may be seen projecting dorsally from the roof of the diencephalon as a slender, club-shaped stalk. The pineal body seems not to function as a light-sensitive "third eye" (as it does in certain reptiles and lower vertebrates); its apparent function is rather that of an endocrine organ.

   Anterior to the pineal body, a portion of the diencephalon may be covered by a network of fine blood vessels known as the (anterior) **choroid plexus**. This network facilitates the exchange of oxygen, nutrients, and waste products across the blood-brain barrier.
4. If you have removed the brain, you may now locate the **pituitary gland** on its ventral surface. The pituitary is subdivided into anterior, posterior, and intermediate lobes, of which the anterior lobe is the largest. The **anterior lobe** secretes the important growth hormone that stimulates the body's increase in size during growth. Other anterior lobe hormones include a thyroid-stimulating hormone that indirectly controls metamorphosis, and a gonad-stimulating hormone that controls seasonal mating cycles and other seasonal changes. The **intermediate lobe** secretes a hormone which controls the production of pigment cells (melanocytes) in the skin. The **posterior lobe** probably secretes several hormones; one of these controls blood pressure, water balance, and smooth muscle tension, and is therefore similar in its effects to the mammalian hormone vasopressin. The posterior pituitary is actually part of the **infundibulum**, a funnel-shaped downgrowth from the floor of the diencephalon.
5. The **optic lobes** are nearly spherical in shape and constitute the midbrain or **mesencephalon**, just posterior to the diencephalon. If you removed the brain, you may follow the **optic fiber tracts** anteroventrally from the optic lobes and observe that the fibers cross one another on the ventral side of the brain, forming the **optic chiasma** (just anterior to the pituitary and infundibulum). The function of the optic lobes is to receive and interpret visual impressions, forwarded from the eyes via the optic nerves.

6. The **cerebellum** is small, but should be visible as a slender, transverse ridge on the dorsal side of the brain just posterior to the optic lobes. The functions of the cerebellum include reception of auditory impressions (transmitted from the ear via the acoustic nerve) and probably also postural reflexes and equilibrium.
7. The **medulla oblongata** is the most posterior portion of the brain. The medulla tapers posteriorly and continues directly into the spinal cord. The functions of the medulla include the control of several important coordinated but largely instinctive reflex actions such as those involved in breathing and swallowing.

   The roof of the medulla is very thin and lies in a depression surrounded by thicker portions of the brain wall. This thin part, or **tela choroidea**, is overlain by a (posterior) **choroid plexus**, similar to the one covering the roof of the diencephalon. Like the anterior choroid plexus, the posterior one also facilitates the exchange of oxygen, nutrients, and waste products across the blood-brain barrier.

B. SPINAL CORD

Using bone-crushing shears or pliers, cut away the dorsally projecting portions (neural arches) of several vertebrae and expose the **spinal cord** beneath. Find the dorsal and ventral roots of the ten pairs of spinal nerves. Posterior to the tenth spinal nerve, the spinal cord tapers off as a **filum terminale**.

The spinal cord contains a **central canal** (or **neurocoel**), surrounded by a layer of **gray matter**, and this surrounded in turn by an outer layer of **white matter**. The gray matter of the spinal cord (and of the medulla, too) is organized into four columns: the **somatic sensory** column (dorsalmost), **visceral sensory** column, **visceral motor** column, and **somatic motor** column (ventralmost), in that order.

PERIPHERAL NERVOUS SYSTEM

C. CRANIAL NERVES (for advanced students only)

The frog, like the shark, has ten pairs of cranial nerves, designated by Roman numerals as follows:

i. **Olfactory nerve**, lying along the olfactory lobe of the brain on its ventrolateral surface. The olfactory nerve transmits sensory impulses only, from the nasal epithelium to the olfactory lobe and cerebral hemisphere.

ii. **Optic nerve**, carrying sensory impulses only, from the retina of each eye to the optic lobes. The right and left optic nerves cross each other below the diencephalon, just anterior to the pituitary, forming the **optic chiasma**.

iii. **Oculomotor nerve**, transmitting motor impulses only, to certain extrinsic muscles of the eyeball (muscles Nos. 19–22, Chapter 3).

iv. **Trochlear nerve**, transmitting motor impulses only, to the obliquus superior muscle (No. 23).

v. **Trigeminal nerve**, carrying both sensory impulses from the jaws and face and also motor impulses to the muscles of the mandibular arch (muscles Nos. 25–28, 30, and 32, Chapter 3). Of the three branches or divisions of the trigeminal nerve, the upper two (**ophthalmic** and **maxillary divisions**) are sensory only, while the lower or **mandibular division**, a mixed nerve, is the only one which carries motor as well as sensory fibers.

vi. **Abducens nerve**, carrying motor fibers to the rectus lateralis muscle (No. 24).

vii. **Facial nerve**, carrying motor fibers to the muscles of the hyoid arch (Nos. 29 and 31, Chapter 3) and also sensory fibers to the tongue. This nerve and the next exit from the brain side by side.

viii. **Acoustic** (or **auditory**) nerve, carrying sensory fibers only from the inner ear to the cerebellum and medulla.

ix. **Glossopharyngeal nerve**, carrying sensory impulses from the tongue and pharynx, and also motor impulses to the anterior petrohyoideus muscle (No. 28). This nerve and the vagus leave the brain together, branching away from each other subsequently.

x. **Vagus nerve**, carrying mostly motor but also many sensory fibers to the larynx, pharynx, and most of the viscera. There is also an **accessory nerve** (sometimes considered to represent the eleventh cranial nerve, though it is here merely part of the vagus and not separately distinguishable), which transmits motor impulses to certain muscles (Nos. 39–43, Chapter 3) formerly associated with the gill apparatus.

The hypoglossal nerve is not a cranial nerve in the frog, since it arises as a branch of the first spinal nerve (see below).

D. SPINAL NERVES

The ten pairs of **spinal nerves** (sometimes designated as Nos. 2 through 11) may be traced laterally from the spinal column, on the dorsal side. Notice that each spinal nerve arises from the spinal cord from both **dorsal** (sensory) and **ventral** (motor) **roots**, and each is therefore a *mixed* nerve. Almost immediately, each spinal nerve gives off a **dorsal primary ramus** to the epaxial musculature and a **visceral ramus** to the autonomic nervous system. The majority of each spinal nerve continues as the **ventral primary ramus** or **somatic ramus**.

Further dissection must be done on the ventral side; be especially careful not to damage any of the digestive organs, blood vessels, or urogenital ducts. Beginning students should notice what they can (the brachial and lumbosacral plexuses in particular) without need of additional dissection. Advanced students should find and identify all ten spinal nerves and their major derivative nerves, as follows:

#1. The first spinal nerve gives off certain small branches to nearby muscles, including a branch which contributes to the brachial plexus. Its major offshoot, however, is the **hypoglossal nerve**, running first laterally and then turning anteriorly to carry motor fibers only to the muscles of the tongue and hyoid apparatus (Nos. 33–37, Chapter 3).

#2. This nerve, together with small contributions from Nos. 1 and 3, constitutes the **brachial plexus** supplying the forelimb. After giving off two branches in the shoulder region, the plexus then gives rise to its principal branch, the **brachial nerve**. This enters the arm and eventually gives off further branches, of which the **radial** and **ulnar nerves** are the largest.

#3. Though it gives off a small branch to the brachial plexus, the principal branch of this nerve quickly achieves the lateral body wall, which it serves with both motor and sensory fibers.

#4. This nerve crosses the dorsal side of the body cavity obliquely to achieve the lateral body wall at approximately the level of the ninth or sacral vertebra; it carries both sensory and motor fibers.

#5. Like the previous nerve, this one runs obliquely toward the lateral body wall and its musculature, carrying both motor and sensory fibers.

#6. This nerve also runs obliquely toward the hip region, carrying both sensory and motor fibers.

#7. This is the first nerve in the **lumbosacral** (or **sciatic**) plexus, which includes spinal nerves #7–10. The two major offshoots of nerve #7 are the **iliohypogastric** and **crural nerves**.

#8. This nerve gives off a small branch to the crural nerve (see #7, above), but its major portion fuses with part of nerve #9 to become the **ischiadic nerve** (*n. ischiadicus*, also called *sciatic nerve*).

#9. This nerve gives off several small branches to the vicinity of the cloaca, but its major portion joins with spinal nerve #8 to form the **ischiadic nerve**.

#10. This nerve, smallest of those contributing to the lumbosacral plexus, gives off branches principally in the region of the cloaca.

E. *AUTONOMIC NERVOUS SYSTEM* (for advanced students only)

The **autonomic**, or visceral motor, nervous system arises in the head region from certain cranial nerves (Nos. iii, v, vii, ix, and x) and also from the visceral rami (see above) of the spinal nerves. There are two functionally distinct divisions: the **sympathetic division**, which stimulates muscular activity, accelerates heartbeat, etc.; and the **parasympathetic division**, which represses muscular activity and retards heartbeat but stimulates digestion.

The **sympathetic trunk** may be found on the dorsal wall of the abdominal cavity as a series of white or silvery masses (the **sympathetic ganglia**) between the transverse processes of successive vertebrae. In searching for them, push the other organs aside *gently*, being especially careful not to damage any of the urogenital structures. The sympathetic trunk parallels the systemic arteries and dorsal aorta quite closely, sending many fine branches to all of the major visceral organs (digestive, urinary, genital). Proceeding very carefully, try to find as many of these branches as you can. Try also to observe the connections between the sympathetic trunk and the visceral rami of the spinal nerves.

The **parasympathetic** ganglia are more difficult to locate since they do not form a distinct or continuous trunk. Important parasympathetic ganglia are located near or upon the major viscera. Most parasympathetic nerves are derived from branches of the vagus nerve; other cranial nerves play more modest roles, and the role of the spinal nerves is virtually nonexistent.

F. *SPECIAL SENSE ORGANS*

1. The **nose**: olfaction.

   The nasal passages have already been described in Chapter 4-A. They are lined with a sensory epithelium, the **nasal mucosa** (or **olfactory epithelium**), from which cell bodies have sent out nerve fibers along the olfactory tract toward the brain. The parts of the brain concerned with olfaction include the cerebral hemispheres in addition to the olfactory lobe and tract; all are described above.

2. The **eye**: vision.

   The eyelids and nictitating membrane have been described in Chapter 1; they protect the eye. The transparent membrane in front of the eye is the **cornea**. Behind the cornea is the **aqueous humor**, containing the **ciliary body** and **iris diaphragm**; further to the rear is the **lens**. The space behind the lens is filled with **vitreous humor**, and the rear wall of the eyeball consists of a double-layered **retina** surrounded on the outside by a **choroid coat** and a **scleroid** (**sclerotic**) **coat**, in that order. The impulse then travels from the retina to the optic nerve, optic tract, and the optic lobes.

The muscles of the eyeball (Nos. 19–24) have been described in Chapter 3.

3. The **ear**: hearing.

The **tympanum**, or eardum, has been described in Chapter 1. Beneath this membrane lies the **middle ear** cavity, connected to the pharynx by means of the **eustacian tube** (Chapter 4-A). The **columella** or **stapes** lies within the middle ear cavity; it was described in Chapter 2-B.

The **inner ear** is the principal organ of both hearing and balance. The footplate, or median end, of the stapes thrusts against a fluid-filled **perilymphatic cistern**, which in turn is connected to the **perilymphatic duct** and **perilymphatic sac**, all filled with **perilymph**. A closely associated system of chambers and ducts are filled with another fluid, **endolymph**; these include the **sacculus, utriculus**, and **lagena**. From the utriculus arise three **semicircular canals**, each with a bulblike swelling (**ampulla**) at one end. The sacculus and lagena are primarily concerned with hearing, the utriculus and its semicircular canals with balance. Concentrations of sensory nerve cells, known as **maculae**, occur in the sacculus, utriculus, and lagena, and in the ampulla of each semicircular canal. One such macula is found in amphibians only and is therefore called **papilla amphibiorum**.

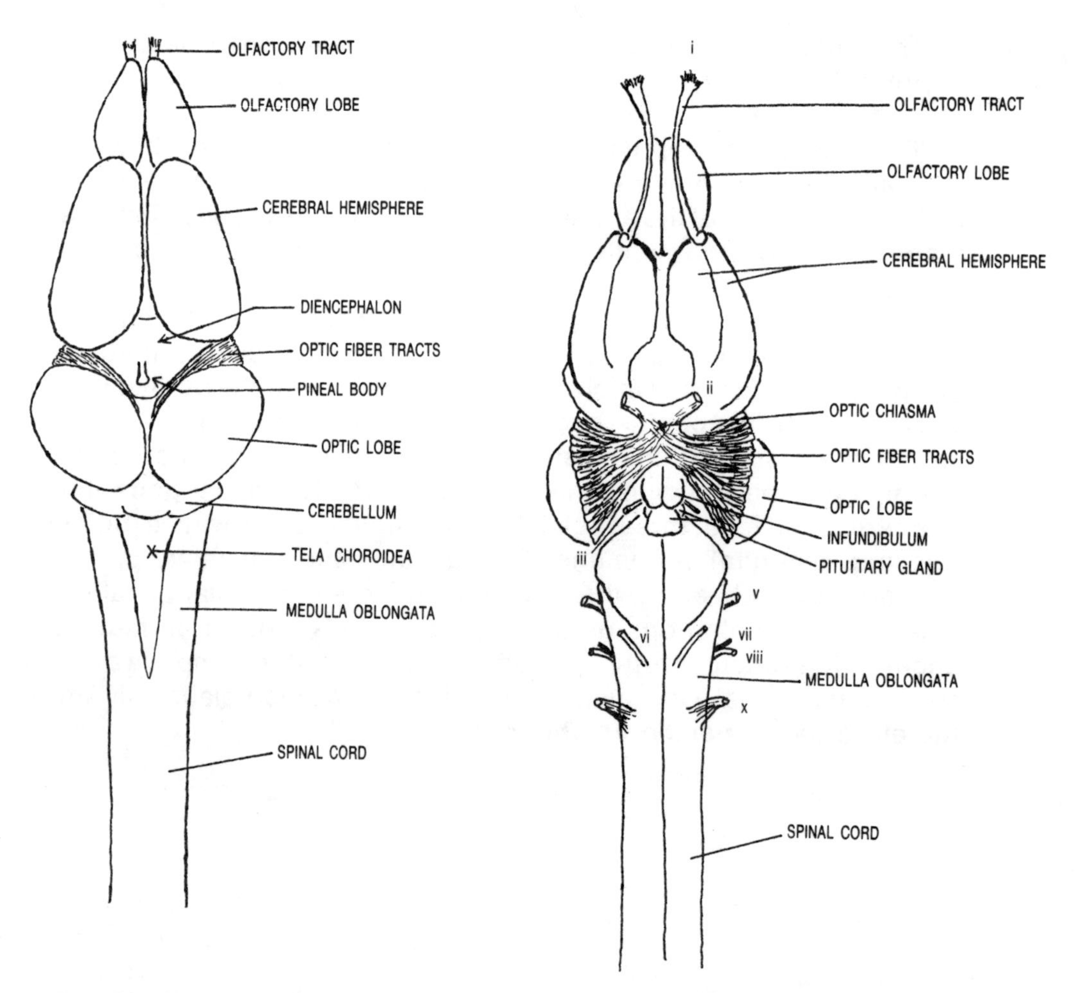

Fig. 37 Brain, dorsal view.

Fig. 38 Brain, ventral view. Roman numerals indicate cranial nerves.

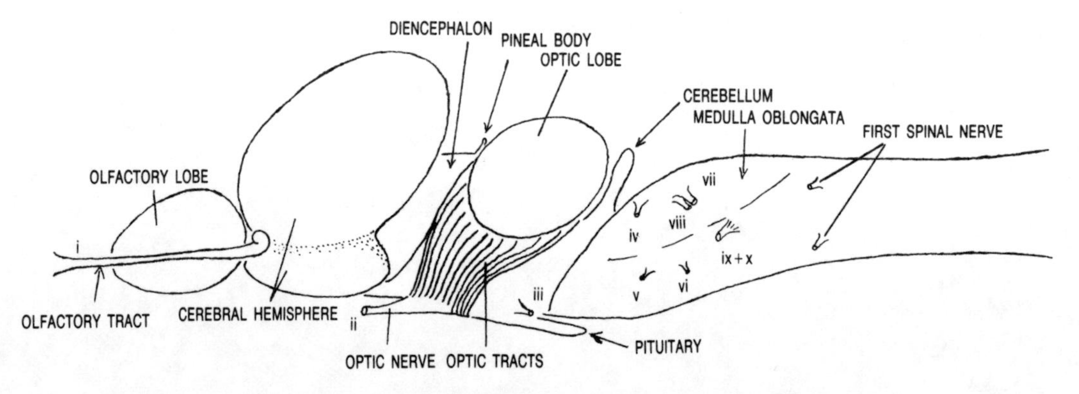

Fig. 39 Brain, lateral view. Roman numerals indicate cranial nerves.

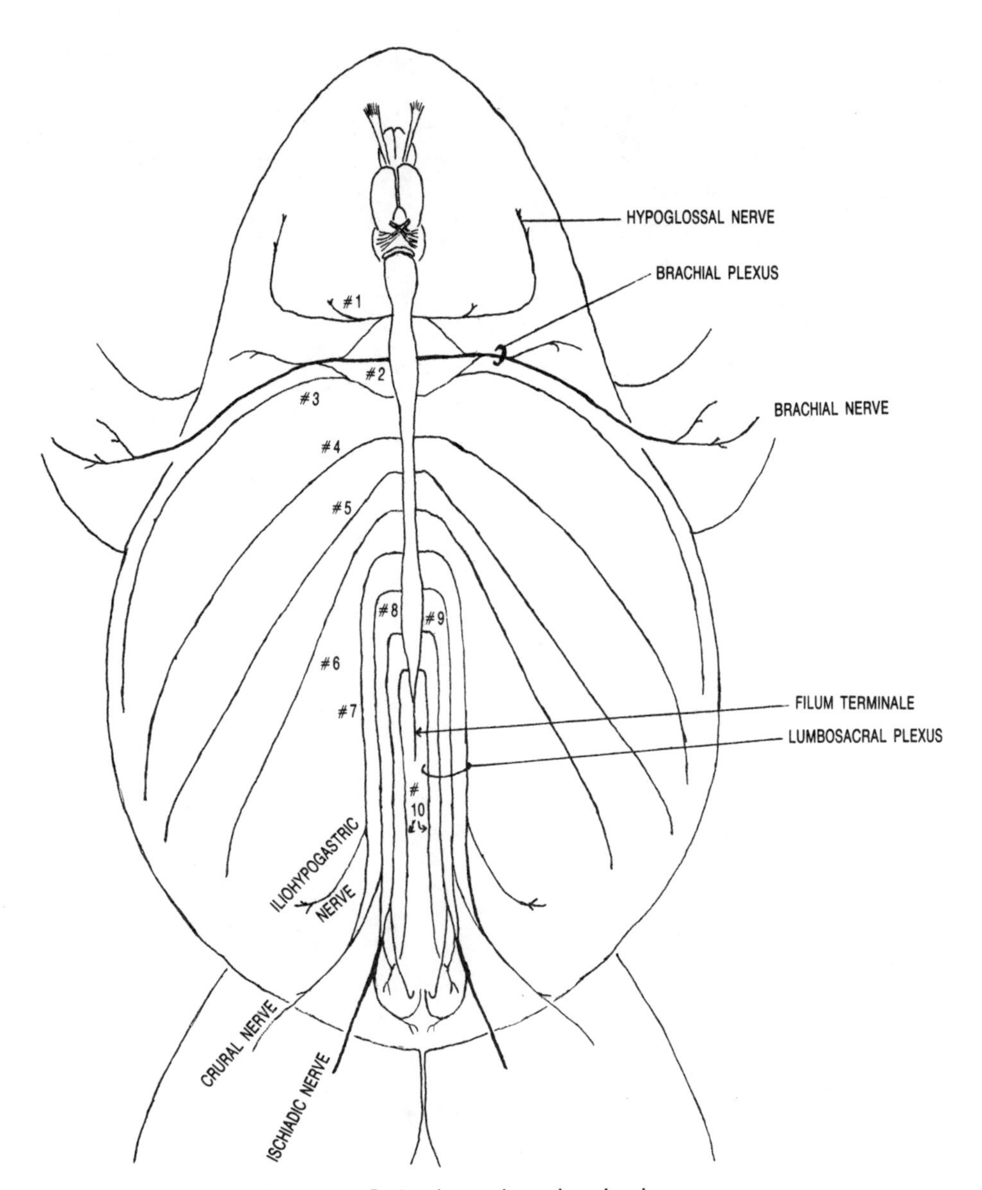

Fig. 40 Spinal cord and spinal nerves.

For Further Reading

Students and others seeking information or detail beyond the scope of this manual will probably find what they are looking for in one of the following reference works:

Ecker, A. and Wiedersheim, R. (1888–1904)
Anatomie des Frosches, zweite Auflage; revised by E. Gaupp.
Braunschweig, F. Vieweg und Sohn.
[The standard reference work on frog anatomy, long out of print.]

Noble, G. K. (1931)
The biology of the Amphibia.
New York, McGraw-Hill; reprinted 1954 by Dover Publications, Inc.
[A classic work on amphibians in general.]

Romer, A. S. (1970)
The vertebrate body, fourth edition.
Philadelphia, London, and Toronto, W. B. Saunders Co.
[One of several widely used textbooks on vertebrate anatomy generally.]

Dunlap, Donald G. (1960)
The comparative myology of the pelvic appendage in the Salientia.
J. Morphol. **106**: 1–76.

Index